Frontispiece:
Shop signs in Cheapside
(*by courtesy of Victoria & Albert Museum*)

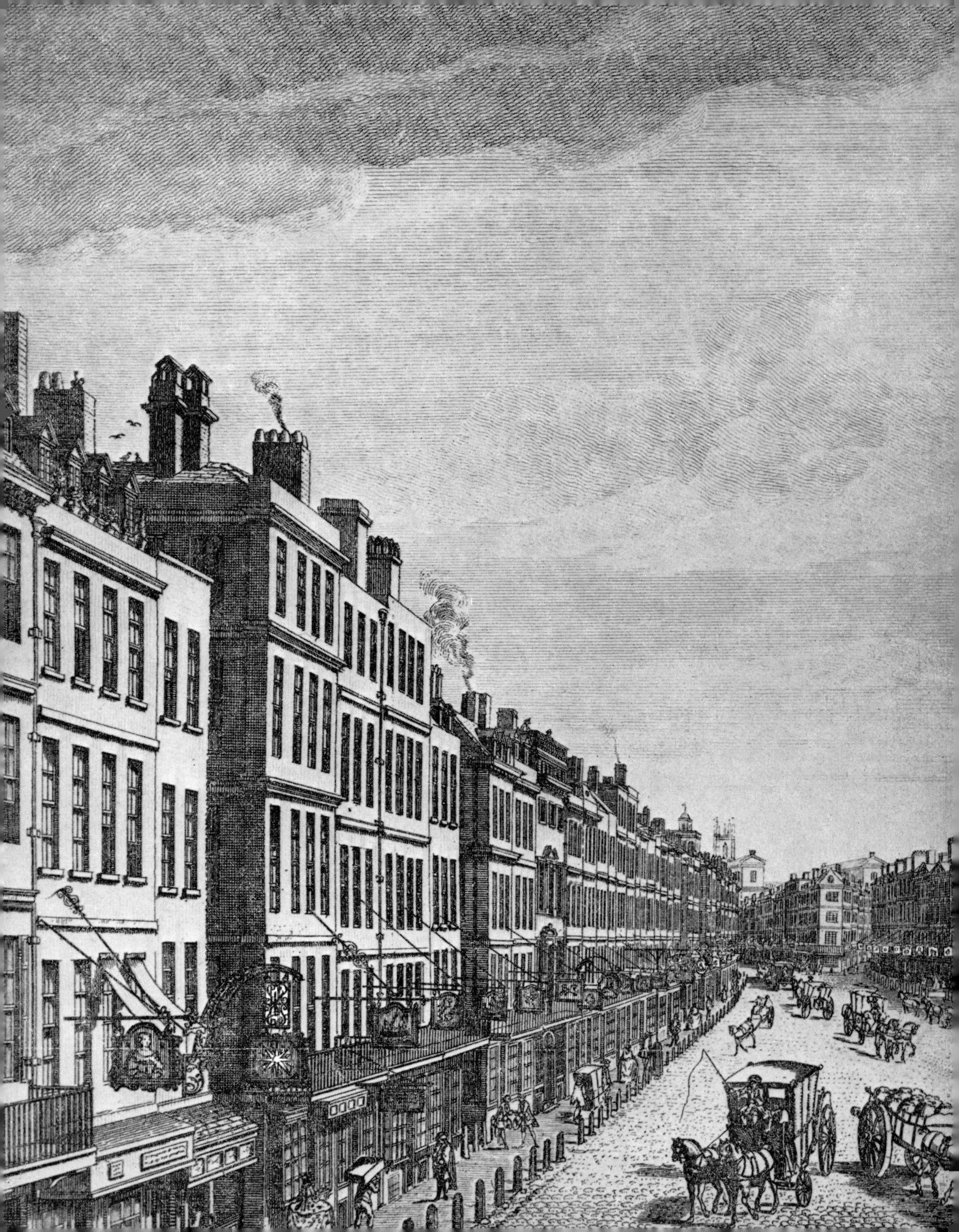

Science Museum

Scientific Trade Cards

in the Science Museum
Collection

by H. R. Calvert M.A.

Her Majesty's Stationery Office London 1971

Printed in England for Her Majesty's Stationery
Office : text by Swindon Press Ltd.
Plates by Artisan Press, Leicester.
Dd 153806 K 16

List of Plates

Plates/Catalogue Number

Frontispiece: Shop signs in
Cheapside, London

1 Dudley Adams 5

2 George Adams 6

3 J. Andrews and Sons 12

4 Attributed to Ayscough 17

5 Fr. J. Berg 45

6 James Ayscough 18

7 James Ayscough 21

8 Thomas Barnett 28

9 John Bennett 41

10 John Bennett 43

11 John Bleuler 52

12 John Bleuler 53

13 William Brind 61

14 John Browne 64

15 Philip Carpenter 71

16 Chadburn Brothers 83

17 Benjamin Cole 91

18 Colombi 94

19 Fisher Combes 95

20 Oliver Combes 96

21 John Cuff 101

22 E. Culpeper 106

23 W. Dowling 132

24 John Gilbert 156

25 Walter Hayes 182

26 A. Hilger 193

27 Nathaniel Hill 194

28 Samuel Johnson 209

29 W. and S. Jones 219

30 W. and S. Jones 221

Plates/Catalogue Number

31 George Lee and Son 233

32 Joseph Linnell 239

33 James Mann 251

34 Benjamin Martin 258

35 Edward Nairne 281

36 Newton and Co. 288

37 J. D. Potter 302

38 Henry Pyefinch 309

39 Ramsden 311

40 Manuel Recarte 314

41 Thomas Ribright 316

42 Matthew Richardson 317

43 R. Rust 334

44 Edward Scarlett 339

45 Samuel Scatliff 340

46 Jonathan Sisson 359

47 Christopher Stedman 381

48 A. Syeds and Co. 391

49 Anthony Thompson 395

50 John Troughton 402

51 J. and E. Troughton 404

52 Thomas Tuttell 407

53 Thomas Tuttell 408

54 Thomas Tuttell 410

55 John Walsh Walsh 424

56 J. Watkins 427

57 John Yarwell 463

58 Samuel Whitford 444

59 Thomas Wright 455

60 John Yarwell 460

61 John Webb 431

Scientific Trade Cards

Definition and history of Trade Cards

Tradesmen's Cards or, more shortly, Trade Cards are essentially advertisements printed on one side of a card or, more usually, a single sheet of paper. They give the name and address of the tradesman, some account of the services and wares he has to offer and perhaps the date of establishment of the business and the name of the predecessor. They vary in shape, and the size may be anything from two or three inches across to a little more than a foot.

The earliest known English trade card was printed shortly before 1630 but they did not become common until about 1720 and the following fifty years saw their heyday. The earliest in the Science Museum's collection is that of Walter Hayes (Plate 25) printed in script type about 1650. Although the earliest ones were plain printed statements, it later became the practice to ornament the cards and to show pictures of some of the things offered for sale. By the middle of the eighteenth century the designs were very elaborate and often included a representation of the shop sign of the premises to help the customer in finding them. Matthew Richardson for example (Plate 42) uses the sign of Sir Isaac Newton's Head and Golden Spectacles. The practice of numbering the premises in London streets began in 1762 and was fairly general by about 1780. Before this time the shops were identified by the signs projecting from the front of them (frontispiece).

Besides being advertisements, given to prospective customers, the trade cards also served as convenient pieces of paper to enter a memorandum or bill of account on the blank back of the card, and this developed into a practice of having an illustrated bill-head with a blank space below (plates 29 and 30). The same block or engraving could also be used to insert an advertisement in a book or newspaper.

The cards continued to be used into the nineteenth century but gradually became less picturesque until they were replaced by the visiting cards used today, which name little more than the firm and the representative making the visit.

Members of all trades used the cards but those listed in this catalogue are almost all of a scientific nature. They are mostly the cards of the makers and sellers of scientific instruments, but also include those of pharmacists, chemists, oculists, photographers and others following pursuits of interest in science.

Use and interest of Scientific Trade Cards

Scientific Trade Cards are useful source material for the history of science and scientific instruments especially if they can be dated. Some cards, such as that of

Thomas Wright (plate 59), bear a date, but there are various means of dating the others. The numbering of premises in the streets, and the introduction of postal districts of London constitute landmarks.

The engraver of the plate puts his signature to it and the date of the engraver may be better known than that of the instrument-maker. The Grainger who signed Bleuler's card (plate 11) was W. Grainger who worked from 1784 to 1793. He illustrated an edition of Bunyan's *Pilgrim's Progress*. James Smith, who engraved Benjamin Cole's card (plate 17) flourished about 1733.

Some of the cards are decorated with the Royal Coat of Arms (eg plates 41, 44, 48 and 61) which can often be dated. In the catalogue the names of the royal, and other, patrons have been included to help with the dating. For instance, John Bennett (plate 10) was instrument-maker to the Duke of Gloucester and Duke of Cumberland. William Henry, son of the Prince of Wales, was created Duke of Gloucester in 1764 and William Augustus, Duke of Cumberland, son of George II, died in 1765, so that the date of this card is determined to within a year (if the information on the card is reliable and as long as Bennett is not claiming to be instrument-maker to somebody already dead).

Once the cards have been dated it is possible to follow the chronological development of the instrument trade, to assess the dates of introduction of the instruments depicted on the cards and to find which maker influenced which. The relative situations of their premises is also material.

A. Syeds & Co claim on their card (plate 48) to have invented the boat's betticle— an early form of the word binnacle, a housing for a ship's compass.

An ornate engraved plate which shows instruments made by James Ayscough (plate 6) also appears in a very slightly amended version (item 19) in which a nautical quadrant replaces two lenses. It looks as if Ayscough at some time decided to add quadrants to his products.

When Joseph Linnell (plate 32) succeeded James Ayscough he took his predecessor's card (plate 7) as it stood and altered only the name.

The illustrations of instruments on Tuttell's card (plate 52) are also used with almost the same lay-out in Fisher Combes's card (plate 19). Tuttell's name has however been removed.

The cards disclose how spectacle-makers hawked their wares from house to house. They left a card or pamphlet describing the spectacles which concluded (item 278) : 'Mr Morris will call with a large assortment . . . This bill will be called for'. The circulars used by the different hawkers (items 120, 278, 279, 366) resemble one another more or less closely in their phraseology.

An indication of the international aspect of the scientific instrument trade is afforded by Tuttell's card (plate 52) written in both French and English and by one of Scarlett's (plate 44) written in Dutch, French and English.

Most of the cards in the collection are those of English makers, predominantly from London, but there are also advertisements of businesses in America, Austria, France, Germany, Italy, Spain, Sweden and Switzerland.

Provenance and nature of the Science Museum collection

In the main the material described in this catalogue came from two collections made originally by Mr Thomas H. Court and Mr George H. Gabb respectively. For about 30 years until the Second World War they were well-known collectors of early scientific instruments and of the relevant trade cards as well as advertisements cut from early newspapers and books. Book illustrations sometimes incorporate an undisguised advertisement for the maker of the instrument illustrated (plate 49) or the instruments may be depicted with the maker's name discreetly showing (plate 53). The instruments shown in plate 53 are all numbered and the item was probably used for a priced catalogue of Tuttell's products.

The items in the collection are of varying merit but it has been thought desirable for the sake of completeness to include them all in the catalogue. To indicate the relative importance of the items they have been divided into four categories. Those in category (1) are authentic Trade Cards and Bill-heads. Category (2) contains useful and informative advertisements, many of them newspaper cuttings. Category (3) includes the second-rate material and photographic reproductions of originals not in the collection, and the items in Category (4) are of little or no scientific interest.

Mr Court pasted the items of his collection into scrap albums. Unfortunately he did not usually record the source of an advertisement cutting, but sometimes he wrote a date on it in manuscript. These dates are recorded in the catalogue.

Many instrument-makers stuck their trade cards—or alternatively, specially-made small labels giving their name and address—inside the lids of the instrument boxes. Some of these have been removed from the boxes and transferred to the scrap albums, but as this usually damages the label, many have been left glued to the boxes in their original position. This fact is indicated in the list of reference numbers at the end of the catalogue by giving the number of the box in which the item is to be found.

Other collections of Trade Cards and bibliography

There are several collections of Trade Cards in the British Museum, Print Room, but their subjects range over all trades not only those of scientific interest. The largest of these collections is that made by Sir Ambrose Heal, which includes about 120 cards of scientific interest. He wrote several papers and books about his collection such as: *London Tradesmen's Cards of the XVIII Century: An account of their origin and use* (1925) and *London Furniture Makers from the Restoration to the Victorian Era, 1660–1840* (1953). His books give useful information about trade cards and are richly illustrated with reproductions of them.

Other collections in the British Museum are those of Banks (with more than 4000 items), Fielden and Alexander Franks.

Mr R. S. Whipple wrote an article on John Yarwell's trade card (item 462) in *Annals of Science* in 1951 (Volume 7, page 62).

Arrangement of the catalogue entries

The words used in the catalogue entries have to a large extent been copied from the original card using the same (sometimes unusual) spelling, punctuation and capital letters. (It has not been thought necessary to use *sic*.) The arrangement of the words has however been altered to accord with the following scheme :

> Name, predecessor, address, date. (The date being at the right-hand end of the line.)
> Trade, patrons, date of establishment of business.

The author's comments on the card are separated by a small space from the quotations taken from the card itself.

At the bottom left of the entry the number in brackets, (1), (2), (3) or (4), indicates the category of the card as defined above. It is followed by size, shape and condition. Sizes are given in inches to the nearest eighth of an inch, vertical dimension followed by horizontal. When the item is other than rectangular (e.g. oval) the largest vertical and horizontal dimensions are given.

At the end of the catalogue there is a list of the Science Museum's reference numbers for the items.

Catalogue

1 **A. Abraham & Co.,** 20, Lord Street, Liverpool. 1855
Opticians and Mathematical Instrument Makers. Established 1817.
(2) $3\frac{1}{2}$ x $4\frac{5}{8}$

2 **J. Abraham,** 7, Bartlett Street, Bath & at His Shop adjoining Mr Thompson's Pump Room, Cheltenham. *c*1837
Optician and Mathematical Instrument Maker to His R.H. the Duke of Gloucester and His Grace the Duke of Wellington.
(1) 3 x $4\frac{1}{2}$

3 **Abraham,** Corner of an Opening into Old Well Lane near Queen's Circus [Cheltenham].
Optician.
(3) $3\frac{1}{8}$ x $3\frac{3}{8}$

4 **Abraham & Dancer,** 13, Cross Street, King Street, Manchester. 1842
Manufacturers of optical, mathematical and philosophical instruments.
(2) $3\frac{1}{2}$ x $4\frac{1}{4}$

5 **Dudley Adams,** Son of George Adams Senior and brother to the late George Adams, 60, Fleet Street, London.
All sorts of Optical, Mathematical & Philosophical instruments constructed.
(1) 16 x $10\frac{1}{2}$ *Plate 1.*

6 **George Adams,** at Tycho Brahe's Head in Fleet Street, London.
Mathematical instrument-maker to His Majesty.
(1) $8\frac{3}{8}$ x 11 *Plate 2.*

7 **George Adams,** at Tycho Brahe's Head, the Corner of Racquet-Court in Fleet-Street, London. 1775
Mathematical Instrument-Maker to His Majesty's Office of Ordnance.
(3) Reproduction $4\frac{1}{4}$ x $6\frac{7}{8}$

8 **George Adams,** at Tycho Brahe's Head in Fleet-Street, London. 1751–60
Azimuth and Steering Compasses Invented By Gowin Knight F.R.S. and made by George Adams Mathematical Instrument Maker to His Royal Highness the Prince of Wales.
(3) Photograph. Oval 3 x $4\frac{1}{8}$

9 **Humphry Adamson** at the House of Jonas Moor Esq. in the Tower makes the instruments described in Samuel Morland's *The description and use of two arithmetick instruments.* 1673
(3) Photograph $5\frac{1}{2}$ x $3\frac{1}{4}$

10 **Robert Akenhead,** at the Bible and Crown upon the Bridge, Newcastle. 1718
Stationer and seller of instruments and spectacles.
(3) $2\frac{1}{2}$ x $2\frac{7}{8}$

11 **J. Alefounder,** Market Street, Faversham. 19th Century.
Chemist & Druggist.
(2) Bill-head. 4 x 7

Elias Allin, *see* John Tomson. Item 399

F. B. Anderson, *see* Stebbing & Wood. Item 380

12 **J. Andrews & Sons,** No. 4 Burlington
Arcade, Piccadilly & at No. 24 Anderson's
Building, City Road.   1820
An advertisement for an exhibition of fancy
glass working, spinning, blowing, variegating, twisting & linking glass.
(3) $8\frac{7}{8}$ x $7\frac{3}{8}$ *Plate 3*

13 **Alfred Apps,** 433, Strand, London, W.C.
near Charing X Railway Station.
Optical, Mathematical & Philosophical In-
trument Maker.
(1) $2\frac{5}{8}$ x $3\frac{3}{4}$

14 **D'Artaria et Compagnie,** Rûe Kohlmarkt
No. 1151, Vienna.
Map shop (French language).
(3) $2\frac{1}{8}$ x $3\frac{1}{2}$

15 **Aslin,** 15, Lower Smith Street, Northamp-
ton Square, Clerkenwell.
Shagreen & morocco case maker.
(2) $1\frac{7}{8}$ x 3

16 **Aubert & Linton,** 252, Regent Street, near
Oxford St. W. London.
Chronometer, Watch and Clock makers
from Geneva, established 1825. Sole repre-
sentatives of the late M. Aubert of Regent
Street. Correspondents to Raingo Frères,
Paris & to H. Capt, Geneva.
(1) 3 x $4\frac{1}{2}$

17 **(Ayscough)**   *c*1740
There is no wording on this illustration of a
pair of spectacles. Mr. Court attributed it to
Ayscough and dated it tentatively as 1740.
He drew attention to the spring in the bridge
of the spectacles, the earliest known.
(3) $2\frac{1}{2}$ x $3\frac{7}{8}$ *Plate 4*

18 **James Ayscough,** at the Great Golden
Spectacles in Ludgate-Street, near St
Paul's, London, (Removed from Sir Isaac
Newton's Head in the same street.)
Optician.
(1) $13\frac{1}{2}$ x 8   *Plate 6.*

19 **(James Ayscough)**
The pictorial part of item 18 differing slightly
by having a quadrant in place of two lenses.
(1) 6 x $7\frac{1}{4}$

20 **James Ayscough**
The text of item 18 without the picture and
with a few variations and an additional foot-
note about the quality of optical glass.
(1) 7 x $8\frac{3}{8}$

21 **James Ayscough,** at the Great Golden
Spectacles and Quadrant, No. 33, in
Ludgate-Street, near St. Paul's, London,
(The Original Shop for superfine Crown-
Glass Spectacles). Optician.
The card is similar to that of Joseph Linnell.
(1) 12 x $7\frac{1}{4}$ *Plate 7*

22 **James Ayscough** at the Great Golden
Spectacles, in Ludgate Street, London.
Oval design for use in lid of instrument box.
(1) $3\frac{1}{4}$ x $3\frac{5}{8}$

23 **James Ayscough** at the Great Golden
Spectacles, Ludgate Street, London.   1751
An advertisement mainly for the supplying
and fitting of spectacles.
(2) $6\frac{5}{8}$ x 4

24 **James Ayscough,** at the Golden Spectacles and Quadrant, in Ludgate-Street,
London.
Advertisement with similar wording to item
23 with additions.
(2) $7\frac{5}{8}$ x $4\frac{5}{8}$ (printed on both sides)

25 **James Ayscough,** Ludgate-Street,
London.   1754
Description of a telescope adapted to use at
Sea, invented by James Ayscough. (In
English, French & Spanish).
(3) Photograph. $12\frac{3}{8}$ x 7

James Ayscough, *see* James Mann &
James Ayscough, 2 items, 253 & 254.

26 **Bancks,** No. 440, Strand, London. (cf. item 27). *c*1800
Makes & sells all sorts of Mathematical, Optical, & Philosophical Instruments on the most Improved Principles and Lowest Terms. Wholesale & Retail.

Oval design for use in lid of instrument box.
(1) Irregular octagon. $2\frac{3}{8}$ x $3\frac{3}{8}$

Bancks, *see* Kaleidoscope sellers, item 223.

27 **Banks,** No. 441 Strand, London (cf. item 26)

Small oval cartouche in lid of instrument box.
(1) Oval $\frac{5}{8}$ x 1

28 **Thomas Barnett,** (No. 61), Great Tower-Street, London.
Optical, Mathematical and Philosophical Instrument-Maker to His Majesty's Boards of Customs and Excise.
(1) 8 x 7. *Plate 8*

29 **Thomas Barnett,** No. 61, Great Tower Street, London.
Optical, Mathematical & Philosophical, Instrument Maker. To His Majesty's Hon[ble] Boards of Customs & Excise (cf. item 30).
(1) $2\frac{1}{2}$ x $3\frac{5}{8}$

30 **Thomas Barnett,** No. 4, Mores Yard, Old Fish Street, near Doctors Commons Makes and sells all kinds of Barometers, Thermo-meters, Hygrometers etc.

The decorative border is like that of item 29 but the address and wording are different.
(1) $2\frac{1}{2}$ x $3\frac{5}{8}$

31 **W. Barrow & W. Lovelace,** at the Golden-Lion in Saint Martins Le-Grand, London.

Wheels and tools for watch-makers, also artificial magnets. Wm. Barrow removed to Woolton near Prescott, Lancashire, 1756.
(1) $4\frac{7}{8}$ x $6\frac{7}{8}$

32 **Rich**[d] **Barry,** No. 106, Minories, London.
Navigational instruments and maps.
(1) $3\frac{1}{8}$ x $4\frac{5}{8}$

33 **Bastick & Driver,** Late Foreman & Apprentice to Mr. Vandome, Leadenhall Street, Scale Makers. No. 2 Holywell Row, Worship St., Near Finsbury Square.
(1) Corners cut from rectangle, 2 x $3\frac{1}{2}$.

34 **R. B. Bate,** 21, Poultry, London.
Optician, hydrometer, saccharometer, mathematical & philosophical instrument maker to her Majesty's honourable Boards of Excise and Customs.

Mentions agents for Admiralty charts: Messrs. Dawson and Melling, Liverpool; R. Neill, Belfast; J. Braham, Bristol; T. Bennett, Cork; Spears & Co., Dublin; W. C. Cox, Devonport; and G. Stebbing, Portsmouth.
(2) $5\frac{1}{2}$ x $2\frac{7}{8}$

35 **R. B. Bate,** Nos. 20 and 21, Poultry, London.
Agent for the sale of Admiralty Charts &c. Hydrometer & saccharometer maker to the Excise & Customs.

Street view showing premises of 'Bate, Optician'.
(2) 3 x $5\frac{3}{8}$

R. B. Bate, *see* Kaleidoscope sellers, item 223

36 **A. Bazzoni,** No. 23, Kingsgate Street, Holborn.
Maker of dolls.
(4) $8\frac{1}{4}$ x $6\frac{1}{2}$

37 **Thomas Beach,** No. 11 in Digbeth, Birmingham.
Scale maker.
(1) $2\frac{1}{2}$ x 6

38 **John Beale,** No. 76, Maid Lane near the
Borough.
Mathematical, Optical and Philosophical
Instrument Maker.
(1) $2\frac{1}{2}$ x $3\frac{5}{8}$

39 **Mr Beard,** Portrait Gallery, Ducie Place,
Manchester, at the back of the Exchange.
Also 17, Wharf Road, City Road, London.
1842.
Photographer.
(3) $3\frac{7}{8}$ x $3\frac{5}{8}$

40 **Beaufoy & Co.,** South Lambeth, London.
Makers of Chloride of Soda according to the
the formula of M. Labarraque of Paris.
A label for a quart bottle.
(3) $6\frac{1}{4}$ x $10\frac{1}{2}$

C. Beilby, *see* Kaleidoscope sellers.
item 223

41 **John Bennett** at the Globe in Crown Court,
between St. Ann's Soho & Golden Square,
London.
Mathematical, Philosophical and Optical
Instruments.
(1) $11\frac{1}{4}$ x $8\frac{1}{8}$ *Plate 9*

42 **John Bennett,** At the Globe in Crown
Court, between St. Ann's Soho and Golden
Square, London.
Mathematical, Philosophical and Optical,
Instruments.
(1) $6\frac{3}{8}$ x $4\frac{1}{2}$

43 **John Bennett,** At the Globe in Crown
Court, St. Ann's Soho, London.
Mathematical Philosophical and Optical
Instruments.
Instrument maker to Their Royal Highnesses
the Duke of Gloucester and Duke of
Cumberland. Also the New Nautical Al-
manack by Authority of ye Commissioners
of Longitude.
(1) $8\frac{3}{4}$ x $6\frac{1}{8}$ *Plate 10*

44 **J. F. Bennett & Co.,** Ellipses Works,
Wright's Hill, London Rd, Sheffield; for-
merly 35, Cemetery Road, Sheffield. 1887
A two page pamphlet on the 'Marvel'
Geometric Compass patented and made by
J. F. Bennett.
(2) $10\frac{1}{4}$ x 8

45 **Fr. J. Berg,** Stockholm, Kammakaregatan,
19. (hörnet af Drottninggatan) Sweden.
Mathematical and Geodetical Instruments.
Established 1850
(1) 3 x 5   *Plate 5*

Berge, *see* Kaleidoscope sellers, item 223

46 **R. Bernard,** 14, Lebanon St., Walworth,
S.E.   *c*1860
Writer & Divider, Thermometers & Baro-
meters.
Hand-written visiting card.
(1) 2 x 3

47 **Bestall,** 1, Victoria Cottages, Royal Road,
Kennington Park, near Beresford St.
Maker of polarizing apparatus, objects &c.
1870
Painter of Diagrams for the Magic Lantern,
&c.
(1) $1\frac{3}{8}$ x 3

48 **N. Bion,** au Quart de Cercle Géométrique,
sur le quay de l'Orloge du Palais, à Paris.
Ingenieur du Roy pour les Instruments de
Mathematique.
Photograph of an original in the Cluny Mus.,
Paris.
(3) Photograph $6\frac{1}{8}$ x 7

Blackburn, *see* Moffett & Blackburn,
item 274

49 **H. G. Blair & Co.,** 95, Bute Street, Cardiff.
1860
Chronometer makers, opticians & compass
adjusters.
From Bristol, est-1829
(1) 3 x $4\frac{1}{2}$

50 **Bleterie,** rue des francs bourgeois, porte St. Michel à Paris.

Bows & arrows & firearms.

(4) $3\frac{1}{8}$ x 4

51 **Bleuler,** 27, Ludgate Street, London.
Optical, mathematical and philosophical instrument maker.

(1) $3\frac{3}{8}$ x $5\frac{1}{4}$

52 **Bleuler,** No. 27, Ludgate Street, London.
Optical, Mathematical & Philosophical In-trument Maker.

Part of the illustration on this card was used by Yeates & Son (item 467) & Field (item 148)

(1) $3\frac{1}{4}$ x $4\frac{3}{4}$ *Plate 11*

53 **Bleuler,** 27, Ludgate Street, London.
Optician

(1) $3\frac{1}{4}$ x 5 *Plate 12*

54 **John Bleuler,** No. 27, Ludgate Street, London. 1790
Optician.

Oval design for use in instrument box.

(1) $1\frac{7}{8}$ x $2\frac{5}{8}$

55 **John Bleuler** (Successor to the late Mr. Whitford) No. 27, Ludgate Street, London.
Optician.

Oval design for instrument box.

(1) $1\frac{3}{4}$ x $2\frac{3}{8}$ (torn & defaced)

56 **Blunt,** No. 22, Cornhill, opposite the Royal Exchange.
Optician, & Mathematical Instrument Maker, to His Majesty.

Circular design for instrument box.

(1) $2\frac{3}{8}$ x $2\frac{1}{4}$

Blunt, *see* Kaleidoscope sellers, item 223

57 **Felice Bosiz** in Milano nella Contr. di S. Radegonda N.986
Ointment.

(4) $2\frac{3}{8}$ x $3\frac{1}{2}$

58 **George Bradford,** No. 99, Minories, near Tower Hill, London.
Real Maker of Mathematical Instruments. Flags, sextants & telescopes.

(1) $2\frac{3}{8}$ x $3\frac{1}{2}$

59 **Timothy Brandreth and George Willdey,** late Servants to Mr. Yarwell and Mr. Stirrop, at the Archimedes and Globe in Ludgate-Street, the corner next St. Paul's Church. 1706
... for some years have made the best sort of Spectacles, that They, and Mr. Marshall, and the rest of the Trade, have sold in their Shops.

cf. items 445 & 446
Photograph of card for Optical and other instruments.

(3) $7\frac{5}{8}$ x 5

60 **Timothy Brandreth** and **George Willdey** at the Archimedes and Globe in Ludgate-street, the corner next St. Paul's. 1706
Similar pictorial heading to item 59 but some variations in the text.

Photograph of a card. *cf.* items 445 & 446.

(3) 8 x $5\frac{3}{4}$

61 **William Brind** (from Mr Read's) at the Hand & Scales in Carey Lane in Foster Lane, Cheapside, London.
Scale Maker.

(1) $6\frac{1}{2}$ x $4\frac{1}{2}$ *Plate 13*

Brookes, *see* Johnston, Brookes ,Hector & Davidson, item 211

62 **James Brown,** Glasgow.
Inventor and maker of Hydrostatical Glass Bubbles for Proving Spirits.

A circular calibration chart for specific gravity.

(3) 3 x 4

63 **John Browne,** at the Sphear & Sun Diall in the Great Minories neare Aldgate London. 1671

An illustration of two sectors.

(2) 7 x $10\frac{1}{2}$

64 **John Browne,** near Wapping old Stairs London.
Compass Maker and Ship Chandler.
(1) $7\frac{5}{8}$ x 5   *Plate 14*

65 **John Browning,** City: 111, Minories, E. Factory: 6, Vine Street, E.C.   1870.
Optical & Physical Instrument Maker to Her Majesty's Government, The Royal Observatory, Kew Observatory &c.
Prize medal 1862. Established 100 years.
(1) $2\frac{3}{8}$ x $3\frac{1}{2}$

Browning, *see* Spencer, Browning & Co. items 369 to 374

66 **J. Bullwinkle,** 61, Leadenhall Street.
Book-binder, Stationer & Gilder.
(4) $2\frac{3}{8}$ x $3\frac{1}{2}$

67 **Byrne & Bishop,** No. 8, Butcher-Row, St. George's Market, St. George's Fields, Boro' Southwark London.
Makers of Machines for Publicans.
(4) $5\frac{1}{4}$ x $3\frac{1}{8}$

68 **John Cail,** 61, Pilgrim Street, and 45, Quayside, Newcastle.   1844
Mathematical and Philosophical Instrument Maker, Optician, &c.
(2) $1\frac{7}{8}$ x $4\frac{1}{2}$

69 **Cameron,** 54, South Castle Street, near the Customs, Liverpool.
Chronometer & Watch Maker.
(1) $2\frac{7}{8}$ x $3\frac{1}{4}$ (damaged)

70 **Gio. Batta Campi,** Genoa.   *c*1750
Optician.
Photograph of an entry in a Swiss sale catalogue.
(3) Photograph $6\frac{1}{2}$ x $6\frac{1}{4}$

H. Capt, *see* Aubert & Linton, item 16

71 **Philip Carpenter,** 24, Regent Street, 4 doors from Piccadilly.
Optician.
Advertises a display of objects under the microscope.
(2) $10\frac{7}{8}$ x $7\frac{7}{8}$ *Plate 15*

Philip Carpenter, Birmingham, *see* Kaleidoscope sellers, item 223.

72 **Carpenter & Westley,** No. 111, New Street, Birmingham (& Carpenter, 24, Regent St. London.)
Opticians.
(1) 2 x 3

73 **Carpenter & Westley,** 24, Regent Street, London, S.W.
Opticians.
A part cut from a larger card shows an engraving of their premises.
(1) Cut to $2\frac{7}{8}$ x $1\frac{7}{8}$

74 **Carpenter & Westley,** 24, Regent Street, London.
Opticians.
Advertisement for a magic-lantern entertainment.
(3) $8\frac{1}{4}$ x $5\frac{1}{4}$ Folded pamphlet, 4 pages.

75 **Carpenter & Westley.**
Programme for a magic-lantern entertainment held at Windsor Castle, April 7th, 1865.
(3) $8\frac{1}{8}$ x $5\frac{1}{4}$ Folded programme, 3 pages.

76 **Isaac Carver,** at the Sign of the Globe-Dial in Horslydown.
Mathematical instrument maker.
Cutting from *Stereometry* by Thos. Everard, 3rd edn. 1696
(2) $2\frac{5}{8}$ x $3\frac{1}{8}$

77 **Jacob Carver,** Son to Isaac Carver, at the Atlas and Quadrant near Cherry-Garden Stairs, on Rotherhith-wall, London.

Makes 'the Sliding Rule mention'd in this Book, and all sorts of Gauging Instruments, useful and proper for His Majesty's Officers of the Customs'.

Below the above entry on the same paper is the advertisement of Tho. Cook, by the Excise Office in the Old Jury. Maker of Gauging and Mathematical Instruments.

On the reverse is the advertisement of John Rowley, q.v. item 327.
(2) $3\frac{1}{8}$ x $3\frac{1}{2}$

78 **Cary & Co.** [Henry Porter], 7, Pall Mall, London, S.W. two doors east from Waterloo Place. Late of 181, Strand, W.C.
Opticians & Scientific Instrument Makers to the Admiralty, &c. Opticians to the Royal London Ophthalmic Hospital. Established 1765.
(1) $2\frac{3}{8}$ x $3\frac{5}{8}$

79 **Cary, Porter, Ltd.,** 7, Pall Mall, London, S.W. two doors east from Waterloo Place, Telephone No. 6736 Gerrard. Late of 181, Strand, W.C. (*See* item 80.)
Opticians & Scientific Instrument Makers to the Admiralty &c. Established 1765.
(1) $2\frac{1}{2}$ x $3\frac{1}{2}$

80 **Cary, Porter, Ltd.,** 7, Pall Mall, London, S.W.
This card differs from item 79 in the telephone number which is now No. 1189 Regent, and there is an addition:
'Works: Finsbury Park, N.'
(1) $2\frac{1}{2}$ x $3\frac{3}{4}$

81 **Casket Portrait Comp.,** 40, Charing Cross (Near the Admiralty)   1864
Special type of photograph. The ten guinea Crystal Cube statuette miniature. Mentions Swain's patent.
(3) $2\frac{1}{4}$ x $4\frac{3}{8}$

82 **C. H. Chadburn,** 71, Lord Street, Liverpool and Albion Works, Nursery Street, Sheffield.
                    1855
Optician and Instrument Maker to H.R.H. Prince Albert. Honourable mention at Exhibition, 1851.
(2) 3 x $4\frac{1}{8}$

83 **Chadburn Brothers,** Albion Works, Nursery Street, Sheffield. Branch Establishment, 71, Lord Street, Liverpool.
Optical Mathematical and Philosophical Instrument Makers to H.R.H. Prince Albert.
(1) $13\frac{3}{8}$ x $16\frac{3}{8}$ *Plate 16*

84 **Chadburn Brothers.,** Albion Works, Nursery-Street, Sheffield.   1842
Manufacturers of Electrical, Electro-Magnetic, and all kinds of Philosophical Instruments and Apparatus. Opticians.
(2) $2\frac{1}{4}$ x $3\frac{5}{8}$

85 **Chevalier,** 158, Palais Royal, Paris.
Optician.
'Ancienne Maison Chevalier Ingénieur de père en fils depuis plus d'un siècle. Le Docteur Arthur Chevalier, Opticien, Officier d'Académie, Seul Successeur, Médaille d'argent Exposition de 1878.'

Lozenge shaped label for instrument box.
(1) $1\frac{5}{8}$ x $2\frac{1}{4}$

86 **Charles Chevalier,** son and sole successor of Vincent Chevalier. Palais Royal No. 163, Factory: Cour des Fontaines No. 1 Bis.
                    After 1844
Ingénieur Opticien.

Awards and medals ranging from 1827 to 1844 are mentioned.
(1) $2\frac{5}{8}$ x $3\frac{5}{8}$

87 **Charles Chevalier,** Palais-Royal, 158, Fabrique et Magasins,
cour des Fontaines, 1bis.
Ingénieur-constructeur d'instruments pour les sciences.
(2) $5\frac{1}{2}$ x $3\frac{3}{8}$

Vincent Chevalier, *see* Richebourg, items 318 & 319

88 **P. Clark,** late Alexander Clark & Co., 7, Nelson Terrace, City Road, nearly opposite Sidney Street, London.
Engineer and blowing-machine manufacturer.
(3) $8\frac{3}{4}$ x $5\frac{1}{2}$ Two copies.

89 **E. M. Clarke,** 428, Strand, London.
Optician and Magnetician.

Three different diagrams on one engraving, intended to be cut up to be stuck in instrument boxes.
(1) Engraved plate $3\frac{7}{8}$ x $2\frac{5}{8}$ Three copies.

90 **B. Cole,** successor to Mr Wright.   1763
Advertisement for orreries cut from the 9th edition of Joseph Harris's *The Description and Use of the Globes and the Orrery.* Names four of Mr Wright's customers.
(2) $5\frac{7}{8}$ x $3\frac{3}{8}$

91 **Benja[n] Cole** at ye Royal Exchange or at his House in Ball Alley, going out of George-Yard, into Lombard-Street.
Maker of the Grand Orrery and all other Mathematical Instruments.

An engraving of the Grand Orrery.
See item 92.
(1) $7\frac{1}{2}$ x $5\frac{1}{8}$ *Plate 17*

92 **Benjamin Cole,** At the Orrery next the Globe Tavern in Fleet Street, London.
Mathematical and Optical Instruments of all Sorts.

The card shows the Grand Orrery almost exactly as in item 91 with the addition of the words : 'The Orrery Made by Cole and Son' which appear partly above and partly below the ecliptic ring.
(1) $7\frac{1}{2}$ x $5\frac{3}{8}$

93 **Collot Frères,** Succ[rs] de Mr. Hirbect, 41 Rue de l'Ecole de Médecine, 41, formerly Rue des Mauvais Garçons, Paris.
Scale maker.
(1) $2\frac{1}{8}$ x $3\frac{3}{4}$

94 **Colombi à Brest et à St. Malo.**
Fabrique d'Instrumens d'Optique de Mathématiques &[a]
Opticien de la Marine de S.A. Mgr. Le Prince de Joinville.

Mentions the exhibitions of 1844 & 1849.
(1) $3\frac{1}{8}$ x $5\frac{1}{4}$ *Plate 18*

95 **Fisher Combes** at ye Mariner & Globe in Broad Street, near ye Angel & Crown Tavern, behind ye Royal Exchange, London.
Maketh and Selleth all Sorts of Mathematical Instruments.

The instruments depicted are very similar to those shown in item 407.
(1) $4\frac{1}{2}$ x $6\frac{5}{8}$ *Plate 19*

96 **Oliver Combs,** (from Mr Scarlet) At the Spectacles ye Second House from Essex Street near Temple Bar, London.
Optician.
(1) $8\frac{1}{2}$ x $6\frac{3}{4}$ *Plate 20*

97 Compass variation.

Paper Volvelle. Two compass cards, one rotating above the other.
(4) $6\frac{3}{4}$ x $5\frac{7}{8}$

Tho. Cook, *see* Jacob Carver, item 77

98 **Andrew Cooke,** at the King's-Arms, Holborn-Hill, opposite the Dog's-Head in the Pottage-Pot, near Hatton-Garden, the Sign of the Original Three Flying-Fishes, over the Door, with the Word Cooke upon it. His House No. 6, in Union-Court, where he has lived near 30 Years.
Bugg destroyer
(4) 4 x $6\frac{1}{2}$

99 **William Court & Marmaduke Hodge-son** at the Mariner and Anchor on Little-Tower Hill, London. 1690
Seller of books and Mathematical Instruments & teacher of mathematics.
(2) $4\frac{1}{8}$ x $3\frac{1}{8}$

100 **James Cox,** No. 51, Banner Street, Saint Luke's, London. From No. 5, Barbican.
Working Optician. Manufacturer of Optical, Mathematical, and Philosophical Instruments.
(1) $2\frac{3}{8}$ x $3\frac{1}{2}$

101 **John Cuff.** At the Sign of the Reflecting Microscope and Spectacles, against Serjeant's-Inn Gate in Fleet-Street.
Optician, Spectacle, and Microscope Maker.
(1) $9\frac{1}{2}$ x $6\frac{5}{8}$ *Plate 21*

102 **John Cuff.** At the Reflceting Microscope and Spectacles, against Serjeants Inn Gate Fleet Street, London. 175–
A Bill-head with the first three figures of the date.
(1) $2\frac{5}{8}$ x $7\frac{3}{8}$

103 **John Cuff.** Opposite to Serjeants-Inn, Fleet Street, London.
Optician. Microscope maker.
Instructions for the use of the reflecting microscope (so called because of the concave mirror used to illuminate a transparent object).
(2) $15\frac{1}{4}$ x $9\frac{5}{8}$

104 **John Cuff,** Opposite Serjeant's-Inn Gate, in Fleet-street, London.
Description of the Gregorian Reflecting Telescope as made and Sold by John Cuff.
(2) 11 x 6

105 **E. Culpeper.** At the Old Mathematical-Shop, the Cross-Daggers in Middle-Moore-Fields.
Mathematical and optical instruments made and sold. See item 106.
(3) Photograph, $6\frac{7}{8}$ x 5

106 **E. Culpeper.** This is similar to item 105 without the accompanying text. The name, E. Culpeper, appears on an illustration of a universal ring dial. The engraver's signature reads : 'E. Culpeper Sculp. London'.
(1) $4\frac{1}{2}$ x 5 Two copies. *Plate 22*

107 **Edmund Culpeper,** at the Cross-Daggers in Moor-Fields. 1710
Maker of Mathematical and Optical Instruments.
(2) $4\frac{1}{4}$ x 5

108 **Edmund Culpeper** at the Black and White House, the Sign of the Cross-Daggers, in Middle-Morefields, London. *c*1720
Maker of mathematical, optical and navigational instruments.
(2) 3 x 5 (stained and damaged)

109 **Ed. Culpeper** at the Old Mathematical Shop, The Cross-Daggers in Moore-Fields.
A small booklet entitled *The Discription and Use of a Set of Portable Microscopes Made and Sold by Ed. Culpeper.*
Describes Wilson's screw-barrel microscope.
(3) Booklet 16 pages $3\frac{1}{2}$ x $2\frac{1}{8}$, with folding plate $4\frac{3}{8}$ x $7\frac{3}{4}$

110 **Richard Cushee** at the Globe and Sun between St. Dunstans Church & Chancery Lane and
Tho. Wright Instrument-maker to His Majesty at the Orrery and Globe near Salisbury Court Fleetstreet London. 1768
Globe and map makers and Surveyors.
Plate illustrating a terrestrial and a celestial globe from Joseph Harris's, *The Description and use of the Globes and the Orrery.* 10th edn.
(2) $6\frac{1}{8}$ x $7\frac{1}{4}$

111 **Nathaniel Cutler** near Wapping-Dock. 1631
Teacher of mathematics and seller of mathematical and navigational instruments and books, maps and stationery.
(2) $4\frac{1}{8}$ x $6\frac{3}{8}$

112 **J. P. Cutts,** Sheffield
Manufacturer, of Spectacles, Telescopes, Opera-Glasses &c.

On the reverse is the card of Cutts & Camm, Manufacturers of Table Knives.
(1) 2½ x 3½

113 **J. P. Cutts,** Division Street, Sheffield.
Manufacturer of Mathematical and Nautical Instruments, Barometers, Thermometers &c. By special appointment Optician to Her Majesty.
(1) 3⅛ x 6⅛

114 **J. B. Dancer,** Manchester.
Optician. Inventor of Microscopic Photographs, manufacturer of all kinds of Optical and Mathematical Instruments. By appointment to Her Majesty's Commissioners. Prize Medal 1862.
(1) 1¾ x 4⅛

Dancer, *see* Abraham & Dancer, item 4.

115 **Steph. Davenport,** against the Distillers in High-Holbourn, near Drury-Lane, London. Makes and sells Machines and Instruments for hydraulics, surgery and experimental philosophy.
(3) Photograph 7 x 5½

Davidson, *see* Johnston, Brookes, Hector & Davidson. Item 211

116 **Messrs. Davis,** 65, Bold-street, Liverpool, and 101, High-street, Cheltenham. 1842 Manufacturers of Mathematical, Optical, and Philosophical Instruments.

This advertisement appears on the reverse of that of Mr Beard, item 39.
(2) 3⅞ x 3⅝

117 **Davis' Brothers,** No. 33, New Bond Street, London. Manufactory—209 High Holborn.　　　　　　　1820
Opticians and Mathematical Instrument makers.

Printed on pink paper, wording as item 118.
(2) 2⅝ x 4½

118 **Davis' Brothers,** No. 33, New Bond Street, London. Manufactory—209 High Holborn.　　　　　　　1820
Opticians and Mathematical Instrument makers.

Printed on pink paper, wording as item 117 but printed in a cursive script.
(2) 3⅞ x 4⅜

119 **Davis, (brothers),** 33, New Bond Street, London.
Opticians and Mathematical Instrument makers to H.R.H. Prince Albert.

Picture of a walking-stick-telescope highly approved of by His Royal Highness, And most of the Nobility and Gentry.
(2) 3¾ x 3½

120 **A. Davis,** 29, Warner-street, Portland-place, New Kent road, London.
Spectacle maker By Royal Authority.

A pamphlet giving a description of the wares. At the end : Licensed Hawker, No. 500. This circular will be called for by A.D. who will have an assortment for inspection.
(2) 7½ x 4¾ folded.

121 **I. Deck,** No. 9, Kings Parade, Cambridge.
Practical chemist and mineralogist.
(1) 2⅜ x 3⅝

122 **I. Deck,** 6, Victoria Terrace, Leamington.
Dispensing & Practical Chemist.
(2) Folded pamphlet, 4 pages, 7⅛ x 4⅜

123 **Dent.**

Directions for using Dents Patent Liquid Compass.
(3) 4 x 5 (poor condition)

124 **Edward Derby.** Successor to the Ingenious Mr Joseph Hickman. Union Court facing St. Andrews Church, Holborn, London. Philosophical Instrument Maker.
(1) 3½ x 5

125 **Mr Dillon** in Long-Acre, next Door to the White-Hart. 1722
Optician.

Cutting from Memoirs of Literature, Vol .4.
(2) $2\frac{1}{2}$ x 4

126 **C. W. Dixey.** No. 3, New Bond Street, London.
Optician & Mathematical Instrument Maker to Her Majesty; the King and Queen of Hanover; their Royal Highnesses the Dukes of Sussex and Cambridge; the Princesses Augusta and Sophia; the Duchesses of Kent, Cambridge, and Gloucester; the King of Belgium &c &c. Established 1777
(2) $8\frac{5}{8}$ x 7

127 **C. W. Dixey.** No. 3, New Bond Street, London. May 9th 1840
Optician, Mathematical & Philosophical Instrument Maker.

Bill head with bill for supplying opera glasses, and repairing other glasses.
(1) $5\frac{1}{8}$ x $8\frac{1}{4}$

128 **Dixey,** 21, King's Road, Brighton 1860
Spectacle Maker and Optician.
(2) 6 x $4\frac{1}{4}$

129 **L. Dixey** (Son of Mr G. Dixey, Optician to the Royal Family, Bond Street, London) 62 King's Road, Brighton. 1843
(2) $3\frac{1}{2}$ x $3\frac{3}{4}$

130 **G. Dollond,** 59, St. Paul's Church Yard, London. 183–
Optician to Her Majesty, the Honble. Board of Customs, &c. &c. Manufacturer of all kinds of Optical, Mathematical, Philosophical & Astronomical Instruments.

A bill head.
(1) $3\frac{3}{4}$ x $8\frac{1}{4}$

131 **P. Dollond,** near Exeter Exchange in the Strand, London. 16 May 1763.
Optician to His Majesty and to His Royal Highness the Duke of York.

Bill head and bill for telescope, camera obscura, microscope etc.
(3) Reduced photograph. $2\frac{7}{8}$ x 4. Two copies.

P. & G. Dollond, *see* Kaleidoscope sellers, item 223

132 **W. Dowling,** West Gateway of Lincoln's-Inn, Serle Street, Lincoln's-Inn-Fields, London. 1829
Working optician, Optical, Mathematical & Philosophical Instruments.
(1) $9\frac{1}{4}$ x $7\frac{1}{4}$ *Plate 23*

133 **Dring & Fage,** 20, Tooley Street, London.
Opticians, Mathematical & Philosophical Instrument Manufacturers. Upwards of half century hydrometer makers to the Honble. Board of Excise of United Kingdom.
(1) $3\frac{1}{4}$ x $4\frac{3}{8}$ Two copies.

134. **Dring & Fage,** 20, Tooley Street, Southwark, London.
Hydrometer, Saccharometer, & Gauging Inst^t. Manufacturers, Mathematical Philosophical & Optical Instruments.Upw^{ds}. of 50 Years Hydrometer Makers To His Majesty's Honble. Board of Excise.

'Sikes's Improved Hydrometer' appears prominently at the top of the card.
(1) $3\frac{1}{8}$ x $4\frac{1}{2}$

135 **Dring & Fage,** 19 & 20, Tooley Street, London Bridge.
Makers of Sikes's Hydrometer, Saccharometer & Gauging Instruments to Her Majesty's Revenue of the United Kingdom & Colonies, the Honorable Victualling Board for H.M. Navy.
(1) $3\frac{1}{8}$ x $4\frac{1}{2}$

136 **Dring & Fage,** 19 & 20, Tooley Street, London.
Manufacturers of Sextants, Quadrants, Compasses, Telescopes & every description of Nautical Instruments. Map & Chart Agents.
(1) $2\frac{1}{2}$ x $3\frac{3}{4}$ Three copies. (In the copy,

136 continued

III 110c, the address is 20, Tooley Street, but the plate was amended for the other copies to become 19 & 20 Tooley Street.)

137 **Dring & Fage,** 19 & 20, Tooley Street, London. 18...

Blank calibration card for a saccharometer.
(3) $2\frac{1}{2}$ x 4

138 **Dring & Fage,** 145, Strand W.C.

This card is similar to item 135 but the old address is crossed out and the words 'Removed to 145, Strand, W.C.' are overprinted in red.
(1) 3 x $4\frac{1}{2}$

139 **Dring & Fage,** 56, Stamford St., London, S.E. Telephone No. 11323 Central. Removed from 145, Strand.
Hydrometer, saccharometer, thermometer, scientific & gauging instrument makers to H.M. Customs, Inland Revenue, Colonial Governments &c &c.
(1) $3\frac{1}{8}$ x $4\frac{7}{8}$

140 **Dring & Fage,** 56, Stamford Street, London S.E.

This card is exactly similar to item 139 but the engraved plate has been altered to insert the word Opticians prominently between the Royal Coat of Arms and the list of instruments made.
(1) $3\frac{1}{8}$ x $4\frac{5}{8}$

141 **J. Duboscq,** Paris, 21, rue de l'Odéon, au fond de la cour.
Instruments de précision. Appareils d'optique moderne.

A note extending to both sides of the sheet describes instruments for projecting and demonstrating to a large audience.
(2) $5\frac{5}{8}$ x $3\frac{3}{8}$

142 **J. Duboscq,** Paris, 21, rue de l'Odéon, au fond de la cour.

Advertisement listing the modern optical apparatus available for sale.
(2) $5\frac{1}{2}$ x $3\frac{1}{4}$

143 **John Dunn,** Sign of the Gilded Globe 50, North Hanover Street, Edinburgh.
Optican, Mathematical & Philosophical Instrument maker.
(1) $2\frac{1}{2}$ x $3\frac{5}{8}$

Egerton Smith & Co., *see* Kaleidoscope sellers, item 223

144 **Elliott, Brothers,** successors to Watkins & Hill, No. 30, Strand, London, from 56, Strand, and 5, Charing Cross.   *c*1854
Opticians &c.
(1) $2\frac{7}{8}$ x $4\frac{1}{2}$

145 **Elliott, Brothers,** 449, Strand, London.
Opticians to the Government.
(1) $1\frac{3}{4}$ x $2\frac{7}{8}$

146 **George Elliott,** (No. 19) Terrace, Tottenham-Court-Road, London; From No. 32, Rathbone Place.   *c*1810

Patent water-pumping engines.
Capacity and price listed on back of card.
(2) $4\frac{5}{8}$ x 3

147 **T. Espin,** South, Lincolnshire.
Teacher of the Mathematics.
(1) $3\frac{1}{8}$ x $4\frac{5}{8}$

148 **Field,** late apprentice to Mr Nairne, No. 74 Cornhill, London.
Optical, Mathematical, & Philosophical Instrument Maker.

This card has the same illustration as that of Bleuler, item 52.
(1) $3\frac{1}{8}$ x $4\frac{3}{4}$ (bottom right hand corner missing).

149 **Field.**
Enlarged photograph of an undamaged example of item 148.
(3) Photograph. $4\frac{1}{2}$ x $6\frac{3}{4}$

Samuell Foster, *see* Antonius Thompson, item 396

150 **John Fowler,** at the Globe in Sweeting's-Alley by the Royal-Exchange, London. 1738 Instruments for Gauging, Surveying, Dialling and Mathematicks. Mathematical Books Sold.
(2) $2\frac{1}{2}$ x $3\frac{5}{8}$

151 **Franklin & Co.** No. 20 St. Anns Sqr. Manchester.
Patent Lever Watch Makers Jewellers Opticians Philosophical & Mathematical Instrument Manufacturers.
(1) $1\frac{3}{4}$ x $1\frac{7}{8}$. Label for watch-case, on green paper, corners cut.

152 **Freeman & New,** Leadenhall Street, London.
Scale makers.
Label for scale box; gives weights and values of gold coins.
(2) Cut to 2 x $3\frac{1}{8}$, corners removed. Poor condition.

153 **Peter Frith & Compy.,** Sheffield. Warehouse in London, 5 Bartlett's Buildings Holborn.   1866
Opticians & Manufacturers of Mathematical Instruments &c. &c.
Bill-head with part of bill for telescopes.
(1) $5\frac{1}{4}$ x $8\frac{5}{8}$

154 **B. Gammage,** Son-in-law of the late J. Dicas, No. 17, Clarence Street, Liverpool.
Manufacturer and only proprietor of the patent for Dicas's hydrometer.
Instructions for use of the hydrometer are printed on both sides of the card. Price-list at end.
(2) $4\frac{3}{8}$ x $2\frac{7}{8}$ Printed by Egerton Smith & Co., Lord Street, q.v.

155 **William Gardiner** may be heard of at Richard's Coffee-house within Temple-bar, at Mr John Gardiner's a Peruke-maker, just

without Temple-bar; at Mr Upton's in Sherrard-street near Golden-square, or at Mr Fowke's an Engine-maker, in King-street, Westminster.   1735
Land-surveyor, see E. G. R. Taylor's *Mathematical Practitioners of Hanoverian England.*
(2) $1\frac{3}{4}$ x $3\frac{1}{2}$

William Gardiner, *see* Edward Laurence & William Gardiner, item 230

156 **John Gilbert** at the Mariner, in Postern Row, Tower Hill, London.
Mathematical, Optical & Philosophical Instruments.
(1) $9\frac{1}{4}$ x $6\frac{3}{4}$ *Plate 24*

Wm. & Thos. Gilbert, *see* Kaleidoscope sellers, item 223

157 **W. Gillett,** 6, Bell Yard, Fleet Street.
Spectacle maker.
Advertisement and price list.
(2) $9\frac{1}{2}$ x $7\frac{1}{2}$

158 **Richard Glynn,** at the Hercules and Atlas in Fleet-street, facing Salisbury-Court.
                                        c 1700
Mathematical Instruments, either for Land or Sea, & Apparatus for Experimental Philosophy.
(2) $7\frac{3}{4}$ x $4\frac{3}{4}$

159 **Valentine Gottlieb.** Manufactory near the Bridge, Blackfriars Road. Orders received at Mr Downer's, No. 153, Fleet Street.
Maker of Axeltrees, cranes, presses, ship's logs.
(1) $8\frac{3}{8}$ x $7\frac{3}{8}$

160 **J. Goujon,** Rue du Bac, No. 6, à Paris.
Map seller.
Label with blank rectangle in centre to insert title of map.
(3) $2\frac{5}{8}$ x $3\frac{1}{2}$. Corners cut to make an octagon.

161 **Jean Goujon,** Rue du Bac, No. 6, à Paris.
Map seller.

Label for map. 'Etats-Unis, Sud-ouest' is written in ink by hand in a blank rectangle in the centre of the label.
(3) $2\frac{3}{4} \times 3\frac{3}{4}$. Corners cut to make an octagon.

162 **C. Gould,** Microscope maker.

Engraving of a flea 'magnified with the second power of C. Gould's improved pocket compound microscope'.
(3) $4\frac{5}{8} \times 7\frac{3}{4}$

163 **Grassi & Rampoldi,** No. 22, Grey Street, Newcastle-upon-Tyne, Third shop below the Turk's Head Inn.   1844
Toys and fancy goods.
(4) $7 \times 4\frac{5}{8}$

164 **James Green** at the Spring Clock, in Cullum-Street, Fenchurch Street, London.

Seller of watchmakers' tools and machines.
(2) $3\frac{3}{4} \times 6$

165 **Dr. Edward Green** at his own Lodgings, within the Head of Marline's Wynd, second Turnpike on the Right Hand, first Door of the Stair, where formerly my Lord Primrose lodged. On Tuesdays at the House of David Dewar on the Shore of Leith, On Thursdays at Dalkeith and on Saturdays at Mussleburgh. Oculist.
(3) $8\frac{1}{4} \times 5\frac{7}{8}$

166 **Gregory.**
Optician.
(1) $2\frac{1}{2} \times 1\frac{3}{4}$

167 **Henry Gregory** at ye Azimuth Compass, near ye India-House, in Leaden Hall Street, London.
Mathematical Instruments for Sea or Land.

A printed reproduction of a card.
(3) $2\frac{3}{4} \times 2\frac{1}{8}$

168 **Gregory & Wright,** No. 148 Leadenhall Street, near the India-House, London. G. Wright from Mr Martin's Fleet Street.
Mathematical and Optical Instrument Maker's. Wright's Patent Quintant to 160D.
(1) cut to $2\frac{1}{2} \times 3\frac{5}{8}$

169 **Willm Griffith,** No. 4, Dolphin-Court, Ludgate Hill, London.
Watch-Springer, Liner, & Shagreen-Case Maker.
(1) $7\frac{1}{4} \times 5\frac{1}{2}$

170 **Henry Gyles** of the City of York.
Glass painting for windows, as Armes, Sundyals.
(4) $5\frac{1}{8} \times 3\frac{3}{8}$

171 **Haering** A l'Aigle d'Or, No. 63. Palais du Tribunat, près le Café de Foi.
Optician, maker and seller of instruments for Physics, Optics and Mathematics.
(1) $3 \times 4\frac{3}{8}$

172 **W. H. Halse,** 22, Brunswick Sq. London.
Professor of Medical Galvanism.

A sketch of Halse's portable galvanic apparatus.
(3) $5\frac{3}{4} \times 10\frac{7}{8}$

173 **J. Harris** at his House in Amen–Corner, near Pater-Noster-Row.   1702
Teacher of mathematicks, author of 'Algebra'.
(3) $3 \times 3\frac{3}{8}$

174 **T. Harris & Son,** 52, G^t Russell S^t, Bloomsbury, London.   1820
Opticians.

Small label for spectacle case. Gilt lettering on red. Royal coat-of-arms.
(1) $1\frac{3}{8} \times 1\frac{1}{4}$, top corners cut off

175 **Thomas Harris & Son's,** No. 52, opposite the gates of the British Museum. *c*1852 Opticians to the Royal Family. Established 70 years.
(2) $6\frac{3}{8} \times 4\frac{1}{8}$

176 **Thos. Harris & Son,** 52, Great Russell Street, Opposite the entrance to the British Museum, London.   1839

Opticians to the Royal Family. Established 60 years.
(2) $8\frac{5}{8}$ x $5\frac{3}{8}$

Thomas Harris & Son, *see* Kaleidoscope sellers, item 223

177 **William Harris & Co.** 50 High Holborn corner of Brownlow Street, London.
                                23rd May 1836
Manufacturers of Optical, Mathematical & Philosophical Instruments.
Bill-head with instructions about a telescope.
(1) $5\frac{1}{2}$ x 5 (Engraved plate $2\frac{3}{4}$ x $3\frac{7}{8}$)

178 **William Harris & Co.** 50 High Holborn corner of Brownlow Street, London.
Manufacturers of Optical, Mathematical & Philosophical Instruments.
(1) $2\frac{3}{4}$ x $3\frac{1}{2}$. Two copies in poor condition

179 **W. Harris & Co.** 50 High Holborn, corner of Brownlow Street, London.
Globe maker.
Illustrations of six globes with dimensions and prices. Page from a book.
(2) $5\frac{1}{4}$ x $3\frac{3}{8}$

180 **Tho. Haselden,** near Union-stairs, a little below the Hermitage, London.   1735
Teacher of Mathematicks.
(3) $2\frac{1}{2}$ x $4\frac{3}{8}$

181 **Francis Hauksbee** (the Nephew of the late Mr Hauksbee, deceas'd) in Crane-Court, near Fetter-Lane in Fleetstreet, London.   *c*1725
Maker of pumps and hydrostatic apparatus.
(2) $7\frac{1}{4}$ x 6

182 **Walter Hayes,** at the Crosse Daggers in Moore feilds Neere Bethlem Gate, London.
Mathematical instrument maker.
Printed in script type.
(1) $3\frac{1}{4}$ x $3\frac{1}{4}$ *Plate 25*

183 **Walter Hayes,** at the Crosse Daggers in Moore feilds Neere the Popes Head Tavern London.
Mathematical instrument maker.
This is a tracing of a card similar to item 182 except that the address is altered. The original appeared on the back of the title page of a book by Samuel Foster, 1654.
(3) $3\frac{1}{2}$ x $3\frac{3}{8}$

184 **Walter Hayes** at the Crosse Daggers in Moor Fields, next door to the Popes-head Taveren.   1680
If any Gentleman studious in the Mathematicks, have or shall have occasion, for Instruments thereunto belonging, or Books to shew the use of them, they may be furnished with all sorts, useful both for Sea or Land either in Silver, Brass, or Wood by Walter Hayes . . . and all sorts of Maps, Globes, Sea-Plats and Mathematical Paper, Carpenters Rules, Post and Pocket Dyals for any Latitude, Steel Letters, Figures, Signes, Planets or Aspects. . . .
Below the above advertisement is one for Thomas Salisbury Esq., bookseller.
(2) $4\frac{3}{8}$ x $2\frac{1}{2}$

Walter Hayes, *see* Anthony Thompson, item 395

185 **Tho. Heath** at the Hercules & Globe next door to ye Fountain Tavern in the Strand London.
Globes, Spheres, Weather-glasses, Mathematical Instruments & Books.
(1) $3\frac{1}{4}$ x $4\frac{7}{8}$

186 **Thos Heath** at the Hercules next the Fountain in ye Strand London.
Mathematical instruments.
Illustration of a double level, a theodolite and a universal sundial.
(1) $7\frac{7}{8}$ x $5\frac{1}{8}$ Two copies.

187 **Thos Heath** at the Hercules & Globe, next the Fountain Tavern in the Strand.   1729
Mathematical Instruments with Books of their use.

Illustration of the obverse and reverse of a sector.
(1) $5\frac{1}{8}$ x $7\frac{1}{4}$

188 **Heath and Wing** near Exeter Exchange in the Strand, London.
Mathematical and Philosophical Instruments.

T. Newman uses same card, see item 284
(1) $3\frac{7}{8}$ x $5\frac{1}{8}$

189 **William Heather** at the Navigation Warehouse, 157, Leadenhall Street, London.
Chart and Map-seller. Sextants & Nautical Instruments.
(1) $2\frac{5}{8}$ x $3\frac{5}{8}$

Hector, see Johnston, Brookes, Hector & Davidson, item 211

190 **M. Henry,** 21, Passage Delorme, and Rue de Rivoli, 12, A Paris.   185—
Oculist-Optician.

Bill-head with advertisement and testimonials. English on one side and French on the other.
(1) $9\frac{7}{8}$ x $5\frac{1}{2}$

191 **Henry's** Patent Balance & Guage, 54, Paternoster Row.

Advertisement and calibration tables taken from the lid of a coin balance. One table gives Weights agreeable to Ld Howe's Proclamation at New York. Agents for the balance are : Mr Stamps, Goldsmith, Cheapside, Messrs. Woolley & Hemming & Messrs. Stibbs & Dean, Fish Street Hill.
(2) $1\frac{5}{8}$ x $6\frac{7}{8}$

192 **James Heskett,** No. 13 Sweetings Alley, Roy^l Exchange.
Map, print & chart seller. Globes & mathematical instruments.
(1) $3\frac{3}{4}$ x $2\frac{3}{8}$

193 **A. Hilger,** 204, Stanhope Street, three doors from Mornington Crescent, N.W.
Astronomical and Optical Instrument Maker.
(1) $2\frac{3}{8}$ x $3\frac{5}{8}$ *Plate 26*

194 **Nathaniel Hill** at the Globe & Sun, in Chancery Lane, Fleet Street, London.
Globe Maker, & Engraver. Mathematical Instruments.
(1) 13 x 8 *Plate 27*

195 **Mr. N. Hill** opposite Serjeants Inn Chancery Lane London.
Globe-maker.
(2) $2\frac{3}{4}$ x $6\frac{1}{2}$

Marmaduke Hodgeson, *see* William Court, item 99.

196 **William Holliwell** (from London) No. 14, North Side Salthouse Dock, Liverpool.
Real manufacturer of Mathematical & Optical Instruments for navigation.
(1) $2\frac{7}{8}$ x $4\frac{1}{2}$

197 **Horne, Thornthwaite, and Wood,** (successor to E. Palmer) 123, Newgate Street, London.   June 1846
Chemical and Philosophical Apparatus.
(2) $2\frac{3}{4}$ x 6

198 **Horne & Thornthwaite,** 122, & 123, Newgate Street, London.
Chemical Philosophical & Photographic Instrument Manufacturers. Operative Chemists and Opticians. Medal at 1851 Exhibition.
(1) $3\frac{3}{8}$ x $4\frac{7}{8}$

199 **Horne & Thornthwaite** Nos. 121, 122, 123, Newgate Street, London, E.C.
Photographers. Royal appointment.

1851 Exhibition medal.
(1) $2\frac{3}{8}$ x $3\frac{1}{2}$

200 **John Hudson,** No. 112, Leadenhall Street, near Saint Mary Axe; and at No. 9, Saint Martin's Court, Leicester Square, London.
Maker of spectacles and optical, mathematical, and philosophical instruments and apparatus.
(1) 11 x $7\frac{3}{4}$

201 **J. T. Hudson,** Henrietta Street, Cavendish Square.
Optician and Spectacle Maker.

An advertisement for a pamphlet called *Spectaclaenia* written by Hudson and sold (for one shilling) by Simpkin and Marshall, London.
(3) $8\frac{1}{2}$ x $5\frac{5}{8}$

202 **Hudson & Son,** Greenwich.
Opticians.

Illustrated directions for using a pedometer.
(3) $7\frac{3}{4}$ x $5\frac{1}{8}$

203 **J. Hughes,** 38, Queen Strt Ratcliffe. E. London.
Manufacturer of Sextants & Quadrants Compasses, Telescopes &c.
(1) $2\frac{3}{8}$ x $3\frac{5}{8}$

204 **G. Hund.** Name, of a maker or engraver, on an illustration of various trigonometrical, carpenters' and interest scales.
(3) $3\frac{3}{4}$ x $12\frac{1}{4}$

205 **All^n Hunt,** No. 7, King's Row, Horslydown, Southwark.

Maker of Ship-models.
(1) $2\frac{1}{2}$ x $3\frac{5}{8}$

206 **A. Hurlimann,** successor to Ponthus & Terrode, 6, rue Victor Considérant, Paris.
(Address '13, Passage Dauphine, Paris' is crossed out.)
Maker of precision instruments.
(1) $2\frac{3}{4}$ x $4\frac{1}{4}$ Two copies

Husbands, *see* T. Wright & Husbands, item 454

207 **J. Hynam,** 7, Princes-Square, Wilson-Street, Finsbury.
Maker of Fusees & Congreves (Matches).
(2) $4\frac{1}{2}$ x 5

208 **Henry Isaacs,** No. 60, Lucas Street, Commercial Road St George's East, London.
Optician & spectacle hawker.
(2) Folded sheet $8\frac{3}{8}$ x $5\frac{1}{4}$

209 **Samuel Johnson** (The Oldest Shop), Successor to the late Mr Mann, At the Sign of Sir Isaac Newton and Two Pair of Golden Spectacles, near the West End of St Paul's London.
Maker of spectacles and optical instruments.
(1) $6\frac{1}{2}$ x 7 *Plate 28*

210 **William Johnson,** 188, Tottenham Court Road, London. W.   188–
Manufacturing optician and spectacle maker. Bill head with first three figures of year. Mentions 1862 exhibition.
(1) 7 x $8\frac{1}{4}$ 4 copies

211 **Johnston, Brookes, Hector & Davidson,** successors to the late George Penton, No. 32, New Street Square, Fetter Lane, London.
Brass founders & lamp manufacturers.
(1) $4\frac{1}{4}$ x $5\frac{3}{4}$

212 **D. Jones,** Charing Cross, London.
                    *c*1770–1780
Manuscript instructions from the case of a $2\frac{1}{4}$-inch refracting telescope made by D. Jones.
(4) $5\frac{1}{8}$ x 7

213 **Thomas Jones,** (Pupil of Ramsden) 62, Charing Cross, London.
Astronomical and Philosophical Instrument Maker to His Royal Highness the Duke of Clarence.
(1) $3\frac{3}{4}$ x $2\frac{1}{2}$. Removed from instrument box, and part torn off.

214 **Thomas Jones,** (Pupil of Ramsden).
Astronomical and Philosophical Instrument
Maker.

A torn fragment of card similar to, but not
exactly like, item 213.
(1) $3\frac{3}{8}$ x $2\frac{5}{8}$ (irregular shape).

215 **Thomas Jones,** From 62 Charing Cross,
4, Rupert Street, Coventry Street.
Astronomical, Mathematical, Optical, and
Philosophical Instrument Maker.
(1) $1\frac{7}{8}$ x $3\frac{3}{8}$

216 **Thomas Jones,** No. 13, Panton Street,
Haymarket, London.
Astronomical, Mathematical, Optical, and
Philosophical Instrument Maker.
(1) 2 x $2\frac{3}{8}$ Two copies, (C.C.39 badly torn).

217 **Thomas Jones,** 62, Charing Cross, removed to 4, Rupert Street, Haymarket.
Optician.

Advertisement and price list of telescopes
and other instruments, removed from instrument box.
(2) $3\frac{5}{8}$ x 5

Thomas Jones, *see* Kaleidoscope sellers,
item 223

218 **W. and S. Jones,** No. 30, Opposite
Furnivals Inn, Holborn, London.   1828
Optical and Mathematical Instrument
makers.
(1) Bill head. $2\frac{1}{2}$ x $7\frac{5}{8}$

219 **W. and S. Jones.**   1852
Similar bill head to above item 218 but with
a different date.
(1) $2\frac{1}{2}$ x $7\frac{3}{4}$ *Plate 29*

220 **W. and S. Jones,** No. 30, Lower Holborn,
London
Opticians, &c.

Advertisement for books by the late G.
Adams and a price list of globes and orreries.
(2) 8 x $9\frac{5}{8}$

221 **W. and S. Jones,** at the Archimedes,
No. 30, Lower Holborn, London, (Nearly
opposite Furnival's Inn.)
Philosophical, Mathematical and Optical
Instrument Makers.

A memorandum sheet with picture of
Archimedes and name and address of firm
at the top. There are three copies in the
collection, two have different MS memor-
anda on the blank part, the third has the
lower blank part removed.
(1) $9\frac{1}{2}$ x $5\frac{5}{8}$ & $5\frac{1}{4}$ x $5\frac{5}{8}$
(3) Photograph $10\frac{7}{8}$ x $6\frac{3}{8}$
Plate 30

222 **W. & S. Jones,** No. 30 Lower Holborn,
London.
Opticians, &c.

Small label for instrument box.
(1) 1 x $2\frac{1}{8}$

W. and S. Jones, *see* Kaleidoscope
sellers, item 223

223 **Kaleidoscope sellers**
The Patent Kaleidoscopes, in various forms,
and with all the latest improvements, are
made and sold, in LONDON, By Messrs.
P. & G. Dollond, St Paul's Church-Yard
Messrs. W. & S. Jones, Holborn; Mr. R. B.
Bate, Poultry; Messrs Thomas Harris & Son,
Great Russell Street; Mr Bancks, Strand;
Messrs. Wm. & Thos. Gilbert, Leadenhall
Street; Mr Berge, Piccadilly; Mr Thomas
Jones, Cockspur Street; Mr Blunt, Cornhill;
Mr Schmalcalder, Strand; Messrs Watkins
& Hill, Charing Cross; Mr Smith, Royal
Exchange; and Messrs. Spencer, Browning,
& Rust, Wapping.
at BIRMINGHAM, By Mr Philip Carpenter;
at BRISTOL, By Mr C. Beilby;
at LIVERPOOL, By Messrs Egerton Smith &
Co. and
at EDINBURGH, By Mr John Ruthven.
(2) $4\frac{1}{8}$ x $3\frac{3}{8}$

224 **Messrs Keyzor,** St Giles's Street, Norwich.
                                          1854
Opticians.

Visiting Ipswich until 14th July offers spectacles to view at Mr G. Smyth's, Pastry Cook and Confectioner, Tavern Street, Ipswich.
(2) $4\frac{5}{8}$ x 3

225 **M & M. Keyzor,** No. 16, Tottenham Court Road, London.
Opticians & Manufacturers & hawkers of Spectacles.
(2) $8\frac{1}{2}$ x $6\frac{3}{4}$

226 **R. Kibble,** 6, London Street, Greenwich.
Working goldsmith, jeweller, watch and clock manufacturer.
Advertisement and price list of watches etc.
(2) $7\frac{1}{2}$ x $4\frac{5}{8}$

227 **Gowin Knight**
A certificate for a magnetic compass signed by Gowin Knight, to be stuck in the instrument box.
(3) Photograph. Oval. $3\frac{3}{8}$ x $3\frac{7}{8}$

Gowin Knight, *see* George Adams, item 8

228 **Heber Lands,** in Halford Court in Fen-Church-street, over against Rood-Lane, near the Dial, London.   1694
Teacher of Mathematics.
(3) $4\frac{7}{8}$ x $2\frac{3}{4}$

Batty Langley, *see* Benj. Scott, item 341

229 **Lattré,** rue St Jacques vis-à-vis celle de la Parchemin i.e. No. 20 A Paris.
Engraver.
Title page of a child's atlas showing mathematical and surveying instruments.
(3) $7\frac{5}{8}$ x 5

230 **Edward Laurence and William Gardiner** may be heard of at Mr Jonathan Sisson's, Mathematical Instrument-maker, in the Strand, or at Mr Mead's, a Goldsmith, near Temple-Bar.   1725
Surveyors.
(2) $5\frac{1}{2}$ x $3\frac{3}{4}$

231 **Lavoisier** de l'Academie Royale des Sciences, regisseur des Poudres et Salpêtres de France.
Book plate
(4) $5\frac{3}{8}$ x $3\frac{7}{8}$

232 **Charles Leadbetter,** at the Hand and Pen in Cock-Lane, Shoreditch, London. 1738
Teacher of practical Mathematicks. Author of Astronomical Works: *Uranoscopia* and *Mechanick Dialling.*
(2) $1\frac{1}{2}$ x $3\frac{5}{8}$

233 **George Lee & Son,** Ordnance Row, The Hard, Portsea, and 3, Palmerston Rd, Southsea.
Manufacturers of Mathematical, Optical & Nautical Instruments to the Honble Corporation of the Trinity House & the Admiralty.
(1) 3 x $4\frac{1}{2}$ (corners removed). *Plate 31*

234 **Lenoir,** au Petit Bénéfice, Quai de l'Horloge-du-Palais, n.59.   1830
Optician.
(2) $1\frac{5}{8}$ x $2\frac{3}{8}$

235 **Henry de Leth.** Advertisement in French.
*c*1735
Seller of maps, prints, books & globes.
Pocket globes by Jean Deur.
(2) $7\frac{5}{8}$ x 3

236 **John Lilley & Son,** 9, London Street, Fenchurch Street, London, E.C.   1865
Nautical & Mathematical Instrument Manufacturers to Her Majesty's Royal Navy. Inventors & patentees of the new liquid Steering Compass.
(1) 3 x $4\frac{5}{8}$

237 **John Lilley & Son,** 7, Jamaica Terrace, Limehouse, London.   1865
Nautical and Mathematical Instrument Manufacturers.
Instructions on the care of compass pivots.
(3) $3\frac{3}{8}$ x $6\frac{1}{2}$

238 **Lincoln,** No. 62 Leadenhall Street, London.
Optician.
(1) $5\frac{1}{2}$ x $3\frac{3}{4}$

239 **Joseph Linnell,** successor to the late
Mr James Ayscough, at the Great Golden
Spectacles and Quadrant, No. 33, in
Ludgate-Street, near St. Paul's, London.
Linnell succeeded Ayscough about 1763
and adopted the latter's trade card, c.f.
item 21.
(1) $12\frac{1}{2}$ x $7\frac{1}{4}$
Plate 32

Linton, *see* Aubert & Linton, item 16.

240 **James Long,** Royal Exchange.
Photograph of instructions, in English &
French, in the lid of a telescope box.
(3) 13 x 8

241 **Jos^h Long,** No. 20, Little Tower Street,
London.   1830
Manufacturer of Sikes's Hydrometer &c.
Illustrated advertisement for hydrometers,
saccharometers and guaging rules.
(2) $11\frac{3}{4}$ x $6\frac{7}{8}$. Printed on both sides.

242 **James Love,** at the Rose, No. 12, Hay-
market, nearly opposite the Opera-House.
(Removed from No. 10).   1788
Perfumer's advert. with MS account &
receipt.
(4) 13 x 8

W. Lovelace, *see* W. Barrow & W. Lovelace, item 31.

243 **E. Lutz,** Rue des Noyers, 49, (Bd. St.
Germain), Paris.
Instruments d'Optique à l'usage des sciences.
(1) $2\frac{3}{4}$ x $2\frac{3}{4}$. Removed from an instrument
box.

244 **Ed Lutz,** Rue des Noyers, 49 (Boulevard
Saint-Germain), Paris.
Opticien fabricant des Instruments à l'usage
des Sciences.
(1) $2\frac{5}{8}$ x 4. From an instrument box.

245 **James Lynch** at the Sign of the Royal
Spectacles, Capel Street, Dublin.
Mathematical Philosophical and Optical
Instrument Maker.
(1) Circle 2-inch diameter for instrument
box.

246 **Ed. Mc. Glade,** 166 Gt. Brunswick St.,
Dublin.
Picture Frame Maker.
(4) 3 x $4\frac{1}{4}$

247 **D. Mc.Gregor,** 38, Clyde Place, Glasgow
& 8, William Street, Greenock.
Manufacturer of Nautical, Mathematical &
Optical Instruments, Chronometers &
Watches. *See* item 248.
(1) $2\frac{7}{8}$ x $4\frac{1}{4}$. Stained, corner missing.

248 **D. Mc.Gregor & Co.** 38, 39 & 40 Clyde
Place, Glasgow, & City Bank (. . .) nearly
opposite the Railway (. . .).
Manufacturers of Nautical, Mathematical &
Optical Instruments, Chronometers &
Watches.
Similar to item 247 with minor alterations.
(1) $2\frac{7}{8}$ x $4\frac{3}{8}$. Badly damaged and part
missing.

249 **Thos. Mc.Intosh,** at the Archimedes and
Golden Spectacles, Great Queen Street
Lincoln's Inn Fields London.
Optician. Makes and Sells all sorts of
Optical & Mathematical Instruments.
(1) $8\frac{3}{4}$ x $7\frac{1}{8}$

250 **A. Mackenzie,** No. 15, Cheapside, near St
Pauls Church Yard, London.
Optical, Mathematical, & Philosophical
Instrument Maker.
(1) $3\frac{5}{8}$ x $2\frac{3}{8}$

251 **James Mann,** At the Sign of Sir Isaac Newton, and Two Pair of Golden Spectacles, near the West End of St. Paul's, London. Optician.

A composite 'trade card' made — at some late date — from three separate pieces.
(1) $10\frac{1}{2}$ x $7\frac{1}{4}$
Plate 33

252 **James Mann**   *c*1750

This consists of only the engraved picture of item 251. The words 'James Mann', just below the picture, have been cut in half.
(1) $4\frac{1}{2}$ x $7\frac{1}{2}$

253 **James Mann and James Ayscough,** At the Sign of Sir Isaac Newton, and Two Pair of Golden Spectacles near the West-End of St Paul's, London.
Opticians.

A sheet (pages 15 & 16) from a book. Printed on both sides with a text very similar to that of item 251.
(2) $7\frac{7}{8}$ x $4\frac{7}{8}$

254 **James Mann and James Ayscough** At the Sign of Sir Isaac Newton and Two Pair of Golden Spectacles, near the West-End of St. Paul's, London.   1747

Advertisement cut from the *London Evening Post*, 12 March 1747, describing improvements to the Compound (or Double) Microscope. (Price Six Guineas with a Book of its Use).
(2) $4\frac{1}{2}$ x $3\frac{5}{8}$

255 **John Marshall** at the Archimedes and two Golden Spectacles in Ludgate Street. The Oldest Shop.
Optician. Maker of Optic Glasses to his Majesty.

Text in English, French & German. Original card in Banks Collection.
(3) Photograph $7\frac{5}{8}$ x $5\frac{3}{8}$

256 **John Marshall,** at the Old Archimedes and Spectacles in Ludgate-street, being the second Spectacle-shop from Ludgate, London.
Optician.
(3) Photograph. $8\frac{1}{8}$ x $5\frac{5}{8}$

257 **John Marshall,** at the Sign of the Archimedes and two Golden Spectacles in Ludgate-street, near St Paul's Churchyard, London, the House being new-built. Optician established 1690.

Trade card incorporates text of letter from Edm. Halley, dated Jan 18. 1693,4 expressing approbation by the Royal Society.
(3) Photograph 8 x $5\frac{3}{4}$

Marshal, *see* Timothy Brandreth & George Willdey, item 59.

258 **Benjamin Martin** at his Shop near Crane Court in Fleet Street.
Philosophical Optical and Mathematical Instruments.
(1) 8 x $9\frac{3}{8}$ *Plate 34*

259 **Benjamin Martin,** No. 171, Fleet-street, London.
Philosophical, Optical and Mathematical Instruments.

Directions for the Use of a Reflecting Telescope with a general advertisement at the bottom.
(2) $9\frac{1}{4}$ x $6\frac{7}{8}$

260 **Benjamin Martin,** Chichester.   1735
By the Author, in Chichester, are taught I Writing. . . . II Arithmetic. . . . IX Astronomy and Geography. . . . XVI The Use and Construction of all the most useful mathematical Instruments.

An advertisement page from a book by Martin.
(2) $6\frac{1}{2}$ x $4\frac{5}{8}$

261 **B. Martin**   1742

Price list of microscopes, telescopes and air pump all made by B. Martin, Inventor of the above Microscopes. Also a list of agents where they are sold in Reading, London, Oxford, Cambridge and Sarum.
(2) $7\frac{3}{8} \times 6\frac{1}{8}$

262 **B. Martin,** London

A map of part of the world at the centre of a compass rose.
(3) $6 \times 6\frac{3}{8}$

263 **Benjamin Martin** at the New Invented Visual Glasses Fleet Street, London.
July 24, 1761.

A bill to Mr Gordon for drawing-instruments and telescopes. Signed by M. Martin.
(3) Photograph $6\frac{3}{4} \times 8\frac{3}{4}$

264 **Benjn Martin**

A receipt in manuscript for the amount shown in item 263.
(3) Photograph $3\frac{1}{2} \times 4\frac{1}{2}$

265 **Benjamin Martin** at the New Invented Visual Glasses Fleet Street, London.
March 9, 1761.

A bill to Mr Gorden for telescope & spectacles. Signed by M. Martin.
(3) Photograph $6\frac{3}{8} \times 8\frac{5}{8}$

266 **Benjn Martin**

A receipt in manuscript for the amount shown in item 265.
(3) Photograph $4\frac{1}{2} \times 3\frac{1}{2}$

Martin, jun, *see* Richard Wakefield, item 421

267 **Martinet** aux galleries du Louvre.

An item from a catalogue (in English) for the sale of a Louis XIV Clock inscribed 'Pouilly Inventor Fecit Parisiis' and 'Henricus Martinot Motum Adjunxit.'
(3) $1\frac{3}{4} \times 3\frac{1}{2}$

268 **Thomas Mason,** 11, Essex Bridge, Dublin. Optician. Mathematical & Philosophical Instrument Maker to His Excellency the Lord Lieutenant and the Irish Court. Established AD 1780
(1) $2\frac{1}{4} \times 3\frac{1}{2}$

269 **Thomas H. Mason,** 5 & 6 Dame Street (near the Castle) late 21 Parliament Street, Dublin. Telephone 3846.   1896
Optician, Mathematical and Philosophical Instrument Maker to His Excellency the Lord Lieutenant and the Irish Court. Established AD 1780

A bill head.
(1) $4 \times 5\frac{5}{8}$

270 **John Mawson,** 9, Mosley Street, New-castle-upon-Tyne.   1860
Retailer of optical instruments etc.
(2) $2\frac{3}{4} \times 3\frac{3}{4}$

271 **Mawson & Swan,** Newcastle on Tyne & London. 33, Soho Square, London.
Photographic apparatus.
(1) $6\frac{3}{8} \times 4\frac{3}{8}$

J. Maxwell, *see* J. Senex and J. Maxwell, item 351.

272 **John Milne & Son,** at the Golden Church Branch Bishops Land, High Street, Edin-burgh.   16 July 1767
Founders and Iron-mongers.

Engraved bill-head on bill to Mr. Mackenzie for work on a sundial.
(1) $6 \times 7\frac{3}{8}$

273 **Pablo (or Paulus) Minguet, Madrid.**
1763
Optical apparatus & spectacles.

Advertisement and directions in Spanish.
(3) Photograph, $11\frac{3}{8} \times 7\frac{1}{2}$

274 **Moffett & Blackburn,** 126, Minories, London.
Scale-makers.
(1) $2\frac{1}{4} \times 3\frac{1}{4}$

275 **H. Moll.** Geographer.   1727
Map of the whole world with the Trade
winds.
(4) $8\frac{1}{4}$ x $10\frac{5}{8}$

276 **Francois van Mols,** Antwerp.
Printer.

An engraved book-plate showing three putti
in a library with microscope, globe, lens and
other instruments.
(3) $3\frac{1}{4}$ x $2\frac{3}{4}$

277 **Francis Morgan,** At the Sign of Archi-
medes and Three Spectacles, No. 27,
Ludgate-Street, near St Paul's, London.
Optical, Philosophical, and Mathematical
Instrument-Maker.

In English & French.
(1) 9 x $5\frac{7}{8}$

278 **G. Morris,** No. 9, Bedford-sq. East
London.
Spectacle maker. Licensed hawker.
(2) $12\frac{5}{8}$ x 10

279 **Samuel Morse,** No. 3, Tenter Terrace,
Prescot Street, London.
Spectacle maker. Licensed hawker.
(2) $13\frac{1}{4}$ x $8\frac{1}{8}$

280 **S. Moss,** 150 (adjoining the Arcade) High
Street, Cheltenham.
Druggist, Operative & Experimental Chymist.
(1) $5\frac{1}{8}$ x $8\frac{1}{2}$ 2 copies

281 **Edward Nairne,** at the Golden Spectacles,
Reflecting Telescope and Hadley's Quad-
rant, in Cornhill, opposite the Royal Ex-
change, London.   c1760
Optical, Philosophical, and Mathematical
Instrument-Maker.
(1) $9\frac{3}{4}$ x $6\frac{1}{2}$
Plate 35

282 **Edward Nairne,** No. 20, in Cornhill,
opposite the Royal-Exchange, London.
Optical, Philosophical, and Mathematical
Instruments.

Directions for using the Achromatic Pros-
pective-Glass, which is illustrated on a
separate sheet.
(2) $11\frac{1}{8}$ x $8\frac{3}{8}$ & 6 x $8\frac{3}{8}$

Nairne, *see* Field, item 148.

283 **Henry Neale** at ye end of St Bartholomew
Lane, near the Royall Exchange, London.
                    17th century.
Scale maker.

Cutting from a sale catalogue describing
scales in a box with maker's label in the lid.
(3) $\frac{1}{2}$ x $3\frac{5}{8}$

284 **T. Newman,** Successor to Heath and Wing,
in Exeter Exchange in the Strand, London.
Mathematical and Philosophical Instruments.
Black lead pencils and Books of the Use of
Instruments.

Similar to Heath & Wing, item 188, with the
addition of the words 'T. Newman Successor
to' at the top and 'near Exeter Exchange' has
been altered to 'in Exeter Exchange'.
(1) $3\frac{3}{8}$ x $4\frac{1}{8}$

285 **Newton,** 66, Chancery Lane.
Globe makers, also orreries, tellurians and
planetariums.

Cutting from a newspaper.
(2) $1\frac{1}{8}$ x $3\frac{1}{8}$

286 **Newton & Son,** 66 Chancery Lane,
London.
Globe makers.

Pictures of 8 types of globes with prices of
various sizes of each type.
(1) $5\frac{7}{8}$ x $3\frac{5}{8}$

287 **Newton & Co.** 3, Fleet Street, London,
Near Temple Bar.
Opticians, Mathematical, Philosophical &
Astronomical Instrument Makers, Globe
Manufacturers, to Her Majesty.

(1) $2\frac{3}{4}$ x $3\frac{3}{4}$. Corners removed, card torn in
two.

288 **Newton & Co.** 3, Fleet Street, London. Near Temple Bar.
Opticians, Mathematical, Philosophical, & Astronomical, Instrument Makers, Globe Manufacturers to Her Majesty.

This is similar to item 287 with the addition of particulars of medals, etc., awarded between 1851 and 1885.
(1) $2\frac{1}{2}$ x $3\frac{5}{8}$. Corners removed.   *Plate 36*

289 **Samuel Newton,** Master of the Mathematical School, Christ's Hospital, Founded by King Charles II.   1710
Author & teacher advertises books printed for Christopher Hussey, at the Flower de Luce in Little Britain.
(3) $4\frac{3}{8}$ x 3

290 **Noseda,** au Palais Royal no. 93.
Opticien.
(1) $5\frac{5}{8}$ x 4

291 **James Oswald,** Saint Martin's Church Yard in the Strand, London. 23 Oct. 1747
Music master.

King George II grants copyright in opera entitled *The Temple of Apollo.*
(4) $10\frac{3}{8}$ x $8\frac{1}{4}$

292 **Thomas Page,** William and Fisher Mount.   1735
A list of about 30 books on navigation, gunnery, mathematics and related subjects, with authors' names, headed 'Books of Navigation Printed for Thomas Page, William and Fisher Mount'.
(3) $5\frac{1}{4}$ x $4\frac{3}{8}$

293 **Thos Parnell,** at the Marine & Quadrant. No. 94, Near the Hermitage Bridge, Lower East Smithfield, London.
Mathematical Instrum^t Maker, & Ship Chandler.
(1) $3\frac{1}{2}$ x $4\frac{1}{8}$

294 **John Patrick** in Ship-Court in the Old-Baily London.
Maker of barometers and thermometers.

A descriptive advertisement giving rules for foreknowing the weather. It is illustrated by a print from a copperplate impressed on the top left-hand quarter of the paper. Letterpress on both sides of the sheet. The two versions differ in the setting of the letterpress and in minor details of the wording.
(2) $9\frac{7}{8}$ x $8\frac{1}{8}$. Engraved plate $5\frac{1}{4}$ x $3\frac{1}{8}$. Two versions.

295 **John Patrick,** over against Bull-Head-Court in Jewen-Street; near Cripplegate-Church.
Maker of weather-glasses.

Directions for fixing the Weather-Glass.
(2) $11\frac{3}{8}$ x $6\frac{7}{8}$

George Penton, *see* Johnston, Brookes, Hector & Davidson, item 211.

296 **S. Phillips,** No. 231, Tottenham Court Road, (nearly opposite Percy Street, late of Rathbone Place).
Spectacle maker. Practical Optician.

Price list of spectacles.
(1) $7\frac{1}{2}$ x 5

297 **Ch. Picquet,** à Paris, Quai de Conti No. 17, entre l'Hôtel des Monnaies et le Pont des Arts.
Géographe ordinaire du Roi et de S.A.R. Monseig^r le Duc d'Orléans.

A label for a map with 'Carte de la Palestin' written by hand in ink in the central blank space.
(3) $2\frac{1}{2}$ x $3\frac{5}{8}$ (corners cut)

298 Planetarium course of Experimental Philosophy. Channel-Row, Westminster.   1728
(3) $1\frac{3}{4}$ x $5\frac{1}{2}$

299 **Francisco Porro,** Gibraltar.
Nautical Instruments.

Bill-head.
(1) $2\frac{7}{8} \times 8$

300 **Henry Porter,** apprentice & successor to
the late W. Cary, 181, Strand, London.
Optician & Mathematical Instrument Maker,
to the Admiralty, War Office, Royal Geographical Society, Christs Hospital, Trinity
House & the Swedish & Norwegian Governments, &c. &c.

Card for instrument box with provision for
recording dates of repair and adjustment and
amount of index error.
(1) $3\frac{3}{8} \times 5$
Five examples in different instrument boxes:
Box 379. Adjusted 9/3/85
Box 103. Adjusted May 1886
Box 111. Adjusted May 1886
Box 384. Adjusted April/85. Index Error $+$
   $10''$.
Box 406. Adjusted April/85.

Henry Porter, *see* Cary & Co., item 78.

Porter, *see* Cary, Porter Ltd., items 79 & 80.

301 **Porter and Hunt (John Hunt),** Poly-
technic Institution, No. 309, Regent Street.
Spectacle makers. Practical opticians.
(2) $8\frac{3}{4} \times 7\frac{3}{8}$

302 **J. D. Potter, successor to R. B. Bate,**
31 Poultry, London.
Hydrometer and Mathematical Instrument
Maker. Chart Agent. Mathematical, Optical
& Philosophical Instruments & Apparatus.
To the Rt. Hon[ble] the Lords Commiss[rs] of
the Admiralty.

There are fifteen copies in the collection,
nine of them still in instrument boxes.
(Only one copy is catalogued here.)
(1) $2\frac{3}{4} \times 4\frac{1}{4}$ *Plate 37*

303 **T. Powell,** 164, Regent Street, late of the
Strand London.   May 1828
Exhibition of Fancy Glass Working to be
given at Mr Brown's, Grocer, Lowgate, next
door to the Mansion-House, Hull. Powell
Established 1811.
(3) $5\frac{1}{8} \times 8\frac{1}{2}$

304 **Powell & Lealand. Ross.**
A manuscript description of a Powell &
Lealand microscope with Ross objectives,
written by W. Kingsley, Cambridge Mathe-
matician, 1852, inventor of the Kingsley
substage condenser.
(4) $8\frac{1}{8} \times 3\frac{1}{4}$

305 **P. Premoli & Co.,** Postern, Newcastle.
Looking Glass, Barometer, & Thermometer
Manufacturers.
(1) $2\frac{3}{8} \times 3\frac{1}{2}$

C. Price, *see* Scott & Price, Seller & Price
and Senex & Price, items 342, 344 & 352.

306 **Andrew Pritchard,** 312, Strand, London,
Near the Church.   1830
Microscope maker, inventor of the Diamond
and Sapphire Microscopes.

Price list of microscopes.
(2) $8\frac{7}{8} \times 5\frac{1}{2}$

307 **Andrew Pritchard,** 162 Fleet Street,
London.
Optician and Spectacle Manufacturer.

List of optical apparatus. Prices of spectacles
and books on microscopic science.
(2) $5\frac{1}{2} \times 8\frac{5}{8}$

308 **John Prujean** living neer New College in
Oxford.
Maker of dials and Instruments for the
Mathematicks.

Directions for use of the Horological Ring-
Dial.
(2) $9\frac{7}{8} \times 5\frac{3}{8}$

309 **Henry Pyefinch,** At the Golden Quadrant, Sun, and Spectacles, No. 67, between Bishopsgate-Street, and the Royal-Exchange, in Cornhill, London.
Optical, Philosophical, and Mathematical Instruments.

A catalogue of about 120 items with their prices.
(1) $12\frac{3}{8}$ x $6\frac{1}{2}$
Plate 38

310 **Quian's** Hydrometer.

Directions for use of the hydrometer.
(3) 7 x $7\frac{7}{8}$

Raingo Frères, *see* Aubert & Linton, item 16.

311 **Ramsden,** Near the Little Theatre in y^e Hay-Market, St. James's.
Mathematical, Optical & Philosophical, Instrument Maker.
(1) $2\frac{1}{2}$ x $3\frac{3}{4}$ *Plate 39*

312 **Aaron Rathbone.**

Photograph of the title page of *The Surveyor* by Aaron Rathbone, 1616.
(4) $7\frac{5}{8}$ x $4\frac{7}{8}$

313 **William Read,** From his House in Durham-Yard in the Strand, London, the Queens-Arms being over the Door.   *c*1701
Oculist. Her Majesty's only Sworn Servant in Ordinary, as also to His late Majesty King William III.
(3) $7\frac{1}{4}$ x $4\frac{1}{2}$.   Text on both sides of sheet.

Read, *see* William Brind, item 61.

314 **Manuel Recarte,** Carrera de S^n Geronimo No. 22. Pral. Madrid.
Surveying instruments.
(1) $2\frac{3}{4}$ x $4\frac{3}{8}$ *Plate 40*

315 **Tho. Reeves.**
Mathematical practitioner. Surveyor. Nativitys Calculated.
(1) $3\frac{3}{4}$ x $5\frac{1}{2}$

316 **Thomas Ribright** at the Golden Spectacles in the Poultry, London.
Optician.

There are two almost identical versions of this card in the collection, one of which has: 'Optician to his Royal Highness George Prince of Wales' and the other 'Optician to his Royal Highness Prince George'.
(1) 9 x $7\frac{1}{2}$ *Plate 41*

317 **Matthew Richardson,** late Apprentice to Mr Scarlett at Sr. Isaac Newton's Head, & Golden Spectacles, against York Buildings in ye Strand.
Optician to his Majesty.
(1) $7\frac{1}{4}$ x $4\frac{3}{4}$ *Plate 42*

318 **Richebourg** No. 69, Quai de l'Horloge à Paris, élève practicien, pendant 10 ans, de Feu Vincent Chevalier, premier Constructeur en France des Microscopes Achromatiques.   1850
(1) $4\frac{7}{8}$ x $7\frac{1}{4}$

319 **Richebourg** No. 69, Quai de l'Horloge, à Paris.
Ingénieur-Opticien. Fabrique d'instruments de précision d'Optique, Physique, Mathématiques, Minéralogie et de Marine. Elève de Feu Vincent Chevalier.

This four-page pamplet uses the same illustration of a collection of instruments as item 318.
(2) $8\frac{5}{8}$ x $5\frac{3}{8}$ Folded pamphlet.

320 **Ripley,** No. 335, Wapping, London. (This printed address has been crossed out and replaced in manuscript by 'Mill Place, Commercial Road, Limehouse'.)

Mathematical, Optical, & Philosophical Instruments, for Sea & Land, Navigation-Books and Sea-Charts.
(1) $5\frac{1}{8}$ x $3\frac{3}{4}$

321 **John Roach,** No. 540 Washington Street, near Montgomery, San Francisco, Cal. 1870
Optician. Thermometers, Barometers. Maker of the Large Transit to Re-Survey the City of San Francisco, 1862.
(1) $4\frac{1}{4}$ x $5\frac{3}{4}$ (Printed in red)

322 **John Roach,** No. 429 Montgomery Street, S.W. Corner of Sacramento, San Francisco, Cal.   1880
Optician. Thermometers, Barometers, &c. Large Transit of 1862.
(1) $3\frac{7}{8}$ x $4\frac{7}{8}$ (Gilt letters on black card)

323 **And^w Ross,** 2, Featherstone Buildings, High Holborn.   1853.
A printed table of magnifying powers of object- and eye-glasses of a microscope. Values inserted in manuscript.
(3) $4\frac{3}{4}$ x $7\frac{1}{4}$

324 **Andrew Ross,** 2 Featherstone Buildings, Holborn.
A manuscript price-list of microscope, objectives and accessories.
(3) $9\frac{7}{8}$ x $7\frac{7}{8}$

325 **Joseph Roux Fils Aîné et Comp.** sur le Port vers St Jean à Marseille.
Navigational and drawing instruments, telescopes, flags and charts.
(3) Photograph, $6\frac{1}{4}$ x $4\frac{3}{4}$

326 **John Rowley,** under St Dunstan's-Church in Fleet-Street.   1702
Mathematical Instrument-Maker.
An advertising testimonial to Rowley signed by John Harris.
(2) $2\frac{3}{4}$ x $3\frac{1}{4}$

327 **John Rowley,** at the Globe under St Dunstan's-Church in Fleetstreet, near Temple-Bar, London. c1720
'Advertisement. The Sliding-Rule, mention'd in this Tract, and all other Mathematical Instruments, particularly the Tobacco-Box mentioned in the Preface are accurately made by Mr John Rowley.
The tobacco-box had a circular slide-rule on the lid. This advertisement is on the reverse of item 77.
(2) $3\frac{1}{8}$ x $3\frac{1}{2}$

328 **Royal Gallery of Practical Science,** Adelaide Street, Lowther Arcade.   1824
Advertisement for an early type of science museum.
(3) $1\frac{7}{8}$ x $2\frac{5}{8}$

329 **Royal Panopticon of Science and Art.** Leicester Square.   1855
Advertisement for an early type of science museum.
(3) $2\frac{7}{8}$ x $3\frac{7}{8}$

330 **Thos. Rubergall,** Coventry Street, Haymark^t, London.
Optician, Mathematical & Philosophical Instrument Maker to his Royal Highness The Duke of Clarence.
(1) $2\frac{1}{2}$ x $3\frac{5}{8}$

331 **Mr George Ruff,** 45, Queen's Road (Exactly opposite the Eye Infirmary) Brighton.   1860
Artist and Photographer.
(3) $2\frac{7}{8}$ x $4\frac{3}{8}$

332 **Mr George Ruff,** 45, Queen's Road, (Exactly opposite the Eye Infirmary) Brighton.
Artist and Photographist.
(3) $4\frac{3}{4}$ x $4\frac{1}{8}$ (Printed on blue paper)

333 **Ruspini,** Pall Mall.   *c*1770
Dentist.

Trade card with newspaper cutting pasted
on back.
(4) $3\frac{1}{8}$ x $4\frac{7}{8}$

334 **R. Rust,** Corner of St Catherines Stairs.
Near the Tower of London. Removed from
ye Minories.
Navigational instruments. Inventor of an
artificial horizon.
(1) 7 x $5\frac{1}{2}$ *Plate 43*

Rust, *see* Spencer, Browning & Co., item
370 & Kaleidoscope sellers, item 223.

John Ruthven, *see* Kaleidoscope sellers,
item 223.

Thomas Salisbury, *see* Walter Hayes,
item 184.

335 **William Salmon Junior**   1733
Illustration of a scale, invented by Salmon,
'for reducing any Module being Divided into
60 minutes . . . to Feet & Inches'.
(3) $6\frac{1}{4}$ x $8\frac{1}{8}$

336 **William Salmon Junior.**   1733
Illustration of a scale, invented by Salmon,
'for the ready Dividing of Modules into
Minutes'. The plate was engraved by
B. Cole.
(3) $5\frac{3}{4}$ x $7\frac{1}{4}$

337 **Thomas C. Sargent,** 2, Thames St, next
to the steam boat landing, Commercial
Docks, Rotherhithe, London. 1800/1820
Mathematical and Optical Instrument Maker.
Navigation books.
(1) $3\frac{3}{8}$ x $4\frac{3}{4}$

338 **I. Sax,** 63, Gray's Inn Lane, Holborn. 1858
Mathematical and Philosophical Instrument
Maker.
(1) $2\frac{1}{2}$ x $3\frac{5}{8}$

339 **Ed^d Scarlett,** at the Archimedes & Globe
near St Ann's Church Soho London.
Optician to his Majesty King George the
Second.

In English, French and Dutch.
(1) $11\frac{3}{8}$ x $9\frac{3}{8}$ *Plate 44*

340 **Samuel Scatliff** at Frier Bacon's Head,
the corner of St Michael's-Alley, Cornhill,
London.
Spectacle, & Optick-Glass Maker.
(1) $5\frac{3}{4}$ x $5\frac{3}{4}$ *Plate 45*

Schmalcalder, *see* Kaleidoscope sellers,
item 223.

341 **Benj. Scott** at the Mariner & Globe over
against Exeter Exchange in the Strand
London.   1725
Mathematical Instrument Maker.

Shows: 'An Inspectional Plain Scale for
Delineating, Architecture, Gardening &c.'.
Invented by Batty Langley of Twickenham,
1725.
(2) $6\frac{3}{8}$ x $14\frac{3}{4}$

342 **B. Scott and C. Price,** at the Mariner and
Globe at Exeter Exchange, in the Strand.
                                        1718
Globes, maps & mathematical instruments.
(2) $3\frac{3}{8}$ x 3

343 **G. Scott,** 4, Butcherhall Lane, Newgate
Street, London.
Mathematical Instrument Maker.
(1) $2\frac{3}{8}$ x $3\frac{1}{2}$

344 **Jere: Seller & Cha: Price.**
Sellers of navigation instruments.
(1) $4\frac{1}{2}$ x $3\frac{1}{4}$ 3 copies.

345 **John Seller**   *c*1671
Hydrographer to the King.

Title page of book: 'Practical Navigation or
An Introduction to the whole Art'.
(3) $6\frac{7}{8}$ x $4\frac{3}{4}$

346 **John Seller** at the Hermitage in Wapping.
Dial maker.
Description and Use of the Cube Dials, very proper to set in any Window, to give the Hour of the Day.
(2) $9\frac{1}{2}$ x $3\frac{1}{2}$

347 **John Senex,** F.R.S. over against St Dunstans Church in Fleetstreet, London.
Prices of maps are given below a picture of a celestial globe.
(1) $7\frac{1}{2}$ x $4\frac{3}{4}$

348 **John Senex,** next the Fleece-Tavern in Cornhil. 1707
Globes, Mathematical Books, Maps and Instruments.
(2) $6\frac{5}{8}$ x $4\frac{3}{8}$

349 **John Senex F.R.S.,** at the Globe, over against St Dunstan's Church in Fleetstreet. 1738
Globe- and map-maker.
(2) $2\frac{3}{8}$ x $5\frac{3}{4}$

350 **Joanne Senex** R.S.S. sub Signo Globi e regione Templi S. Dunstani in Fleetstreet, Londini.
Globe, sphere and map maker.
Latin text below a picture of an armillary sphere.
(2) $5\frac{5}{8}$ x $3\frac{5}{8}$

351 **J. Senex and J. Maxwell,** at the Globe in Salisbury Court. 1714
Globe and map makers.
(2) $4\frac{7}{8}$ x $3\frac{5}{8}$

352 **J. Senex and C. Price,** next the Fleece-Tavern, in Cornhill. 1707
Globe makers.
(2) $4\frac{1}{8}$ x $3\frac{3}{8}$

353 **Mary Senex** widow of the late John Senex F.R.S. at the Globe, over-against St. Dunstan's Church in Fleet-Street.
Globes and maps.
(2) Cut into two pieces: $8\frac{5}{8}$ x $11\frac{1}{8}$ & $8\frac{3}{8}$ x $11\frac{1}{8}$

354 **Barlee Shreeve** in the Cockey-Lane, Norwich.
Haberdasher and cutler.
(4) $13\frac{3}{4}$ x $8\frac{3}{8}$

355 **Hen^y Shuttleworth** at Sr Isaac Newton's Head, No. 23 Ludgate Street, London. NB. The Oldest Shop.
Optician.
A label from an instrument box.
(1) $2\frac{3}{8}$ x $3\frac{1}{4}$, corners removed.

356 **Ch^{les} Simonneau,** A Paris, Rue de la Paix No. 6. Hôtel Mirabeau, vis-à-vis le Timbre.
Engraver. Seller of globes, spheres & maps.
(1) $2\frac{7}{8}$ x $3\frac{5}{8}$

357 **James Simons** At Sir Isaac Newton's Head, the Corner of Marylebone Street opposite Glasshouse Street, London. 1785 Mathematical Philosophical & Optical Instruments.
(1) $7\frac{3}{4}$ x $6\frac{3}{8}$

358 **J. Sisson.** *c*1735
Page from a book of illustrations of instruments made, and some invented, by J. Sisson.
(2) $7\frac{5}{8}$ x $11\frac{1}{2}$

359 **Jonathan Sisson** in the Strand.
Mathematical Instrument Maker to His Royal Highness The Prince of Wales.
(1) $8\frac{1}{2}$ x 6 *Plate 46*

360 **Jonathan Sisson** at the Sphere the Corner of Beaufort-Buildings in the Strand, London.
Mathematical Instrument Maker.
(1) $6\frac{3}{8}$ x $4\frac{3}{8}$
(Duplicate of lower half of this card: 1951–685. V 34).

361 **Jonathan Sisson** At the Corner of Beaufort-Buildings, In the Strand, London. 1737
Mathematical Instrument-Maker.
(2) $7\frac{5}{8}$ x $4\frac{1}{2}$ (Printed on both sides).

Jonathan Sisson, *see* Edward Laurence &
William Gardiner, item 230.

362 **Mr Smeaton,** London.   May 12, 1767
Consulting engineer.
(3) 3 x 4¾

Smith, Royal Exchange, London, & **Egerton
Smith & Co.** Liverpool, *see* Kaleidoscope
sellers, item 223.

363 **J. Smith.** From Fraser's, Bond Street.
17 Bath Place, New Road, Tottenham
Court Road.
Practical optician.   Established 1817
Inventor of the Compound Engraver's
Glasses.
(2) 11¼ x 7⅝

364 **James Smith** at the Brunswick Coffee-
House in Fleet-Street.   1717
Artificial eye-maker and oculist.

Portrait of Smith aged 47, with rhyme below.
(3) 12¾ x 8⅜

365 **John Snart,** 215, Tooley-Street, London.
Optical, Mathematical and Philosophical
Instruments.
(2) 3⅛ x 5½

366 **D. Solomon,** Newgate-Street, Newcastle.
Spectacle manufacturer and working optician.
(1) 2½ x 3½

367 **E. Solomons,** No. 36, Old Bond Street,
Piccadilly; and No. 1, Old Jewry, City.
Optician. Optical, Mathematical and Philosophical Instruments.
(2) 10 x 7¼

368 **J. Somalvico and Co.,** Hatton Garden,
London.
Opticians.

Advertise a walking-stick telescope.
(1) 8⅜ x 10½. Two copies.

369 **Spencer, Browning & Co.,** 111, Minories,
and 6, Vine Street, City, E.C. June 1st, 1862
Wholesale manufacturers of Mathematical,
Philosophical, Meteorological, Optical, Nau-
tical, and Surveying Instruments.
Established 100 years.

Trade catalogue and price list of telescopes.
(2) 7⅝ x 5⅜. Folded pamphlet, 2 pages,
printed on 4 sides.

370 **Spencer, Browning & Co.** late Spencer,
Browning & Rust.

Trade catalogue and price list of quadrants
and sextants.
(2) 8¼ x 5½. Folded pamphlet, 2 pages,
printed on 4 sides.

371 **Spencer, Browning & Co.,** 111 Minories.
Factory at 6, Vine St, America Square, E.C.
Manufacturers of Nautical, Optical & Mathematical Instruments. Wholesale bunting
factors & flag makers. Established 100
years.

Short price list of telescopes. (Captains'
double night glasses) in a cover with a
coloured picture of the factory.
(2) 9 x 5¾

372 **Spencer, Browning & Co.** 111, Minories,
factory—6, Vine Street, America Square,
London, E.C.   July 1862
Manufacturers to the trade of Nautical,
Optical and Surveying Instruments.

Price list of magnetic compasses.
(2) 7⅞ x 4⅞. Folded pamphlet, 2 pages,
printed on 4 sides.

373 **Spencer, Browning & Co.** 111, Minories:
factory: 6, Vine Street, America Square,
London.
Wholesale Manufacturers of Mathematical,
Optical, Nautical and Surveying Instru-
ments.   Established 100 years.

Description and price of the Improved
Pancratic Telescope patented by the firm.
(2) 10 x 8

374 **Spencer, Browning & Co.** 111, Minories, and 6, Vine Street, City, E.C. July 1862 Wholesale Manufacturers of Mathematical, Philosophical, Optical, Nautical and Surveying Instruments. Established 100 years. Patentees of Spectroscope, Pelorus, Improved Telescopes and Improved Aneroid Barometers.

Description, with price, of the patent Long Range Aneroid Barometer.
(2) $9\frac{7}{8}$ x $7\frac{7}{8}$

Spencer, Browning, & Rust, *see* Kaleidoscope sellers, item 223.

375 **Spencer and Perkins,** No. 44, Snow Hill. Watchmakers.
Advertisement for the Improved Pedometer or Waywiser.
(2) $8\frac{7}{8}$ x $5\frac{1}{2}$. Two copies.

376 **Spilsbury,** in Russel Court Covent Garden, London.
Engraver. Map and Printseller.
(3) $2\frac{3}{4}$ x 4

377 **Stanley,** Great Turnstile, London, England.
Mathematical instrument maker.
(1) $3\frac{1}{2}$ x $7\frac{5}{8}$

378 **Joseph Stanley,** 67 near Bell Dock, Wapping – London.
Mathematical Instrument maker, and Ship Chandler.
(1) $3\frac{1}{8}$ x $4\frac{3}{4}$

379 **J. R. & H. Stebbing,** No. 63, High Street Southampton. 1833
Opticians, And Mathematical Instrument Makers and General Manufacturers, patronised by Their Royal Highnesses the Duchess of Kent and Princess Victoria.

On the back of the sheet is an advertisement for medicinal baths in Southampton.
(2) $6\frac{7}{8}$ x 4

380 **Stebbing and Wood,** 47, High Street, (nearly opposite the market), Southampton. Optical, Mathematical and Nautical Instrument Makers, to H.M. the Queen, the Royal Yacht Squadron, &c.

The pamphlet advertises the patent spectacles of F. B. Anderson, optician of Gravesend, with a price list.
(2) $8\frac{7}{8}$ x $5\frac{1}{2}$

381 **Christopher Stedman** At the Globe on London Bridge.
Mathematical Instrument maker.

In one copy (1951–686) the words 'Stedman Fecit' do not appear on the semi-circular protractor. Two copies.
(1) $10\frac{1}{4}$ x $7\frac{3}{4}$ & $7\frac{5}{8}$ x 6 (corners cut).
Plate 47

Stirrop, *see* Timothy Brandreth and George Willdey, item 59.

382 **Egbert Storer,** 7 Upper Barnsbury Street, Islington, London, N.
Horologist.
(1) $3\frac{3}{4}$ x 5

383 **Egbert Storer,** 46, Myddelton Square, Clerkenwell.

Describes a model to illustrate the phenomenon of the moon's rotation.
(3) $9\frac{3}{8}$ x $7\frac{5}{8}$

384 **P. Sullivan** at ye South back of St Clement's Church near Temple Barr London.
Circular label for instrument case.
(1) 1-inch diameter.

385 **Patrick Sullivan** without Temple Barr London.
A paper scale for a mercury barometer.
(3) 5 x $3\frac{3}{8}$ mutilated.

386 Sundial.
Engraving of the plate of a sundial with Latin inscription.
(3) $4\frac{1}{2}$ x $4\frac{3}{8}$

387 **Henry Sutton** in Threadneedle Streete neere the Royall Exchange. London. 1654
Illustration of diagonal scale and protractor, also a magnetic compass card for telling the time of day.
(2) $5\frac{7}{8} \times 3\frac{3}{4}$, mutilated.

388 **H. Sutton.** 1658
Engraving of a Sutton hand quadrant of 4-inch radius.
(3) $4\frac{7}{8} \times 5\frac{1}{4}$

Swan, *see* Mawson & Swan, item 271.

389 **Thos. Swann,** 43, Banister S^t, Marybone, Liverpool.
Mathematical Instrument Maker, Navigational Instruments.
(1) $3\frac{1}{4} \times 3\frac{7}{8}$

390 **Thos. Swann,** No 3 Manns Island, Liverpool.
Hadleys Quadrants & other Instruments.
(1) $6\frac{3}{8} \times 4\frac{1}{8}$

391 **A. Syeds & Co.** 379, Rotherhithe S^t near Kings Stairs.
Mathematical Instrument Makers, To His Majesty's Navy.
Inventor of the Boats Betticle.
(1) $3\frac{5}{8} \times 2\frac{1}{2}$, corners cut. *Plate 48*

392 **John Syeds** At No. 17, near East-Lane Stairs, Rotherhithe. 1794
Navigational instruments including Syed's Patent Quadrant (illustrated).
(1) $5\frac{7}{8} \times 4\frac{1}{8}$

393 **Robert Taylor,** High Market Place, Berwick upon Tweed. 1717
Bookseller, Stationer and Printer selling charts, slide-rules, prospect glasses etc.
(2) $5\frac{3}{8} \times 3\frac{1}{4}$

394 **W. N. Telford,** Bridge Street Row, Chester. 1855
Optician, Watch Maker and Jeweller.
(2) $3\frac{3}{8} \times 4\frac{1}{4}$

395 **Anthony Thompson** in Hosier Lane neare Smithfeild.
Between the engravings of the front and back of a sector is written: 'These and all other Instruments for the Mathematicall practice are made by Walter Hayes at the Crosse daggers in Moore feilds next dore to the Popes head Taverne and allso by Anthony Thompson. . . .'
(2) $4\frac{3}{8} \times 7$ *Plate 49*

396 **Antonius Thompson.**
Illustration of a sector marked: 'Delineavit Antonius Thompson' and, 'This Scheame representeth the Sector according to the last alteration by Mr Samuell Foster'.
(2) $2\frac{3}{8} \times 9\frac{7}{8}$

397 **George Thompson & Son,** Thorner Cotton Mill near Leeds.
Manufacturers of Argands and Lamp Wicks.
(1) $2 \times 2\frac{7}{8}$

Thornthwaite, *see* Horne & Thornthwaite, items 197 to 199.

398 **Ainsworth Thwaites** of Rosoman's-Row, Clerkenwell. 28 August 1764
Clock-maker.
An appeal for donations signed, amongst others, by Justin Vulliamy, Clock-maker, near St James's Gate, Pall-Mall and by John Ellicott, Watch and Clock-maker, at the Royal Exchange, because a fire has destroyed Thwaites's business.
(3) $5\frac{1}{2} \times 5\frac{3}{8}$

399 **John Tomson** in Hosier lane, neere Smithfield in London. 1611
An advertisement cut from A. Hopton's *Topographical Glass*, 1611, reads: 'You may have any of the Instruments in this booke made of wood, in Hosier lane, neere Smithfield in London, by John Tomson. The Glasse is made in brasse, in blacke Horse-ally, neere Fleetbridge, by Elias Allin'.
(2) $1\frac{3}{8} \times 4\frac{3}{8}$

400 Townly, at Walworth, St Mary, Newington, Surrey. 1780
Nurseryman.

Advertises Golden Eye-Water for dropping into inflamed eyes.
(4) $6\frac{3}{4}$ x $5\frac{1}{8}$

401 Edwd Wm Townly, Walworth, Surrey.

A label to show excise duty of three pence has been paid.
(4) Cruciform. Two arms an inch wide and about 5 inches long cross each other at right angles.

402 John Troughton, Successor to Mr Cole, No. 136, Fleet Street, London.
Mathematical, Optical & philosophical Instrument-Maker.
(1) Oval $2\frac{1}{2}$ x $3\frac{1}{2}$. *Plate 50*

403 J. & E. Troughton.

The copper plate of item 402, above, has been amended to substitute 'J. & E.' for 'John', otherwise it is identical.
(1) Oval $2\frac{1}{4}$ x $3\frac{1}{8}$

404 Made by **J. & E. Troughton,** No. 136 Next the Globe Tavern in Fleet Street, London.
Mathematical Instrument Makers.

On this label for an instrument box is written in pencil: '1808 14 Feb. Index Error 5".05 to be added'.
(1) Oval 2 x $3\frac{1}{2}$. *Plate 51*

405 Troughton & Simms, London.
Opticians, and Mathematical Instrument Makers to the Honourable Board of Ordnance.
(1) $1\frac{1}{2}$ x $3\frac{1}{8}$, corners cut.

406 Chas. Tulley.

Two receipted bills made out to Mr John Lecky, dated respectively 1814 (for £27-11-0) and 1817 (for £2-8-9).
(3) Each bill 3 x $7\frac{3}{8}$.

407 Tho. Tuttell at ye Kings Arms & Globe at Charing Cross & against ye Royal Exchange in Cornhill London. 1698–1700
Mathematical Instrument Maker To the Kings most Excellent Majesty. (Text in both English & French).

Design of the card similar to that of Fisher Combes, item 95.
(1) 4 x $6\frac{3}{8}$ *Plate 52*

408 Tho. Tuttell, Charing Cross.
Mathematical Instrum^t maker to ye Kings most excell^t Majesty.

The only clues to the maker are given by the inscriptions on three of the more than 60 instruments depicted. Each illustration is numbered as if the picture went with a catalogue.
(2) Two copies. $7\frac{1}{2}$ x $11\frac{3}{8}$ & 8 x $11\frac{1}{2}$.
Plate 53

409 Tho. Tuttell at the Kings-Arms and Globe at Charing-Cross, and against the Royal Exchange in Cornhill.
Mathematical Instrument Maker to the King's most Excellent Majesty.
Advertises Tuttell's Mathematical Cards containing all the Instruments generally us'd in Navigation, Surveying, Dialling, Gauging, Fortification &c.
(2) $3\frac{5}{8}$ x $3\frac{1}{4}$

410 Tho. Tuttell at the Kings-Arms and Globe at Charing-Cross, and against the Royal Exchange in Cornhill, London. 1698
Mathematical Instrument-maker to the King's most excellent Majesty.
Advertises Globes, Cœlestial and Terrestrial.
(2) $3\frac{1}{8}$ x $3\frac{3}{4}$
Plate 54

411 Tho. Tuttell at the Kings-Arms and Globe at Charing Cross, and against the Royal Exchange in, Cornhill. 1698
Mathematical Instrument-Maker to the Kings most Excellent Majesty.

Advertises an astronomical calculator.
(2) $3\frac{1}{4}$ x $3\frac{1}{4}$

412   **Tho. Tuttell** at the King's Arms and Globe, at Charing Cross.   1698
Mathematical Instrument-maker.

Advertises a newly published book on *The Description and Uses of a New Contriv'd Eliptical or Double Dial.*
(2) $2\frac{1}{4} \times 4\frac{1}{4}$

413   **Richd Vandome & Co.** No. 117, Leadenhall Street, London.
Scale-Makers to His Majestys Public Boards, Mint, Exchequer, Bank of England, & the Honble United East India Company.
(1) $3 \times 4\frac{1}{2}$, mutilated, some lettering obliterated.

414   **Cornelius Varley,** 1, Charles Street, Clarendon Square.   1811
Patentee of the Graphic Telescope. Microscopes &c. Drawing, Optical, Philosophical & Chemical Instruments made to order.
(1) $2\frac{1}{2} \times 3\frac{7}{8}$

415   **Cornelius Varley,** No 228, Tottenham-Court Road.

Advertisement for Patent Graphic Telescope with list of prices. Besides from the inventor and patentee, C. Varley, the instruments may be had of Messrs. Smith, Warner & Co. No. 211, Piccadilly; Mr Ackerman, No. 101, Strand and other Venders of Drawing Materials.
(2) $8\frac{5}{8} \times 5\frac{3}{8}$

416   **Cornelius Varley,** Charles Street, Clarendon Square, Somers Town.   1810
Visiting card.
(3) $1\frac{5}{8} \times 3\frac{1}{8}$

417   **Mr Cornelius Varley,** 1, Charles Street, Clarendon Square.   1811
Artist. Patents drawn and Specified.
(3) $1\frac{5}{8} \times 3\frac{1}{8}$ Two copies

418   **Cornelius Varley,** No. 1, Charles Street, Clarendon Square, London.

Advertisement (printed in copper-plate script) for a series of Drawing Books of Boats, Shipping etc. drawn by the aid of the Graphic Telescope.
(3) $8\frac{5}{8} \times 6\frac{3}{8}$

419   **Varley & Son,** 1, Charles Street, Clarendon Square, London.   c1845
Opticians. Chemical & Philosophical Instrument Makers.
(1) $2\frac{1}{2} \times 3\frac{7}{8}$ Two copies.

420   **C^t Vaslin,** Paris, 44, Rue Mazarine, au 1er étage.
Seul concessionnaire du Stéréoscope Duboscq.
(2) $5\frac{5}{8} \times 3\frac{1}{4}$

421   **Richard Wakefield,** No. 13, Southampton Buildings, Chancery Lane, Holborn.
Inventor of a patented method of tuning harpsichords, piano-fortes and spinnets. 'The Patentee has engaged with Mr Martin, jun. Mathematical Instrument Maker, Fleet-Street, (in consequence of a considerable improvement by him) for the sole makeing and executing his said invention.'
(2) Two sheets. $9 \times 11\frac{1}{4}$ & $8\frac{7}{8} \times 11\frac{5}{8}$.

422   **Mr Walker, jun.** Theatre Haymarket.
Advertisement for lecture demonstration with Eidouranion or large transparent orrery.
(3) $1\frac{3}{8} \times 3\frac{1}{2}$

423   **J. & A. Walker,** No. 47, Bernard Street, London. Navigation & Stationery Warehouse at 33, Bottom of Pool Lane, Liverpool. Geographers & Hydrographers. Maps, Charts, Nautical Instruments.
(1) $2\frac{7}{8} \times 4\frac{1}{2}$

424   **John Walsh Walsh** successor to Mr Samuel Shakespear Soho & Vesta Glass Works, Birmingham Heath, Birmingham.   1855
(1) $3\frac{1}{2} \times 4\frac{3}{4}$ *Plate 55*

425 **Julius Wanschaff,** Berlin S.W., Wartenburg-Strasse 14.
Workshop for mechanical and optical precision instruments.
(1) $2\frac{1}{2}$ x 4

426 **Jeremiah and Walter Watkins,** No. 5, Charing Cross, London.
Optical, Mathematical and Philosophical Instrument Makers to His Royal Highness The Duke of Clarence.

The name of the firm is preceded by the words: 'Repaired by'.
(1) $1\frac{1}{8}$ x $2\frac{7}{8}$

427 **J. Watkins,** No. 5, Charing Cross, London.
Optical, Mathematical & Philosophical Instrument Maker, To their Royal Highnesses The Dukes of York & Clarence, and the Hon^e East India Company, cf. item 428.
(1) $2\frac{3}{8}$ x $3\frac{1}{4}$ *Plate 56*

428 **Watkins & Hill.**
This card has been printed from the same plate as item 427 with the name of the firm changed and a letter 's' added to the word 'Maker'.
(1) $2\frac{3}{8}$ x $3\frac{1}{4}$

429 **Watkins and Hill,** 5, Charing Cross, London.   1842
Optical, Mathematical, Philosophical, and Chemical Instruments and Apparatus.

Advertisement for a priced catalogue of instruments.
(2) $2\frac{1}{8}$ x $3\frac{1}{2}$

Watkins & Hill, Charing Cross, *see* Kaleidoscope sellers, item 223.

430 **J. Webb,** No. 408 Oxford-Street, near Soho Square, London.
Optician. All sorts of Optical, Mathematical & Philosophical Instruments Made & Repair'd.
(1) $4\frac{5}{8}$ x 3

431 **John Webb.** Removed from the Old Established Shop, Oxford Street, to Tottenham Court Road opposite the Chapel, London.
All sorts of Optical, Mathematical, & Philosophical Instruments Made & Repaired.
(1) $8\frac{1}{4}$ x 6
Plate 61

432 **Webb & Co.,** Kilworth Cottage, Alfred Street, Maiden Lane, Battle Bridge.
Manufacturer of the Berzelius Match.

Placard headed: 'Sold Here'.
(3) 11 x 8

433 **Nathaniel Wegg,** New Cross Road, Deptford, (a few Doors from High Street).
Manufacturing chronometer, watch, and clock maker. Established ten years.
(2) $7\frac{3}{8}$ x $4\frac{5}{8}$

434 **Alex^r Wellington** (Successor to the late Mr James Search) At the Globe, Crown Court, St Anns, Soho, London.
Mathematical Instrument-maker to their Royal Highnesses the Dukes of Gloucester & Cumberland.
(1) $3\frac{3}{8}$ x 5.   Two copies.

435 **Mr West.** Clifton (Bristol).
Lecturer invites visitors to Observatory at Clifton where there are astronomical instruments including a Cassegrainian telescope by Tully of London and a Transit Instrument by Troughton. Prices of admission.
(3) $8\frac{1}{8}$ x $6\frac{5}{8}$

436 **West.** Fleet Street, London 9th Nov. 1837
Optician.

Picture of West's shop decorated on the occasion of the visit of Queen Victoria to the City of London—used as an advertisement for West's Treatise on the Eye.
(2) $4\frac{3}{4}$ x $2\frac{7}{8}$

437 **F. West,** 83, Fleet Street, near St Bride's Church.
Philosophical Instrument Maker.

Advertises small air-balloons and instructions for filling them with hydrogen.
(2) $5\frac{1}{2}$ x $8\frac{3}{4}$

438 **Francis West,** (Successor to Mr Adams), 83, Fleet Street, London.
Optician to his late Majesty.
(2) $5\frac{1}{8}$ x $2\frac{7}{8}$

439 **Francis West,** (Successor to Mr. Adams, Optician to His Majesty's) 83, Fleet Street, near St Bride's Church.   *c*1831
(2) $2\frac{7}{8}$ x $5\frac{1}{2}$

440 **F. L. West,** 39, Southampton Street, Strand.
Optician & Instrument Maker.

Small label for spectacle case.
(1) Oval with axes $\frac{7}{8}$ x $1\frac{1}{4}$

Westley, *see* Carpenter & Westley, items 72 to 75.

441 **Westwood,** No. 23, Princes Street, Leicester Square, London.   May 1824
Chronometer, Watch and Clock Maker.
Watches used on Capt. Parry's last Voyage.
(1) $2\frac{1}{2}$ x $3\frac{1}{2}$

442 **M. White.**

'M. White's Sliding Rule' appears on an engraved illustration of a slide-rule.
(3) 2 x $12\frac{3}{8}$

443 **D. M. Whitehouse.** (Many years Manufacturer to Bradbury and Rubergall). 1, Castle-Street, Leicester-Sq.
Spectacle maker.

Advertisement and price list.
(2) $6\frac{3}{4}$ x $7\frac{3}{8}$

444 **Samuel Whitford.** (The Original Shop). At the Three Spectacles, No. 27, Ludgate-Street, near St Paul's, London.
Optical, Philosophical, and Mathematical-Instrument-Maker.
(1) $10\frac{1}{2}$ x $6\frac{3}{4}$
Plate 58

Whitford, *see* John Bleuler, item 55.

445 **Willdey** at the Corner of Ludgate-Street, next St Paul's, London.
Optical instrument maker and retailer of instruments, toys, jewellery and prints.
(1) 11 x $12\frac{1}{4}$

George Willdey, *see* Timothy Brandreth and George Willdey, items 59 and 60.

446 **George Willdey and Timothy Brandreth.** At the Archimedes and Globe, Spectacle and Toy-Shop, next the Dog-Tavern in Ludgate-street; a pair of large Globes being on the Posts before the Door.
(3) Photograph, 8 x $4\frac{1}{2}$

447 **Robert Williamson,** Near the Exchange, Liverpool.
Stationer and Bookseller sells Chirurgical, Optical, Philosophical, Mathematical and Musical Instruments.
(1) $8\frac{3}{8}$ x $5\frac{1}{8}$

448 **(J. Wilson).**

A manuscript describing illustrations of a microscope which J. Wilson inserted in his pamphlet of 1706.
(3) $3\frac{7}{8}$ x $5\frac{1}{4}$

449 **W. Wilton,** St Day, Cornwall.
Mathematical Philosophical & Optical Instrument Maker.
(1) $3\frac{1}{4}$ x $4\frac{7}{8}$, corners cut off.

Wing, *see* Heath and Wing, item 188.

450 **Winter.** Late with Frasers. No. 9, New Bond Street.   1812
Working optician.
(1) $2\frac{1}{8}$ x 3

451 **T. B. Winter,** 55, Grey Street, Corner of High Bridge, Newcastle on Tyne.
Manufacturer of Mathematical, Nautical, Philosophical and Optical Instruments.
Spectacles.
(1) $3\frac{1}{8}$ x $4\frac{1}{2}$

Wood, *see* Horne, Thornthwaite, and Wood, item 197

452 **Edward George Wood,** 117 Cheapside (late of 123, Newgate St) London.
Optician, Manufacturer of Photographic, Chemical, Mathematical, & Philosophical Instruments, & Apparatus.
(1) $2\frac{3}{8}$ x $3\frac{5}{8}$ (corners rounded by cutting)

G. Wright, *see* Gregory & Wright, item 168.

453 **T. Wright** at ye Orrery & Globe in Fleet-street London.
Instrument maker to His Majesty.
Small label for sticking inside instrument box.
(1) $2\frac{1}{8}$ x $2\frac{1}{4}$

454 **T. Wright & Husbands**
Above Wright's label (exactly as item 453), a label has been attached to the same instrument box for 'Husbands, 8, St Augustine's Parade, Bristol'.
Husbands presumably repaired or re-sold an instrument made by Wright.
(3) Photograph, $3\frac{7}{8}$ x $2\frac{5}{8}$

455 **Thomas Wright** at the Orrery & Globe next the Globe & Marlborough Head Tavern in Fleet Street London.   1718
Mathematical Instrum^t Maker to his Royal Highn^s George Prince of Wales.
(1) $6\frac{1}{2}$ x $3\frac{7}{8}$ *Plate 59*

456 **(Thomas Wright).** The Orrery & Globe.   1718
Impression from an old wood-cut block probably used for newspaper advertisements. Design similar to the upper part of item 455.
(3) $5\frac{5}{8}$ x $4\frac{7}{8}$

457 **Thomas Wright** at the Orrery and Globe in Fleet Street London.
Mathematical Instrument Maker to His Majesty.
(3) Photograph, $6\frac{3}{8}$ x $4\frac{5}{8}$

Tho. Wright, *see* Richard Cushee, item 110.

458 **Mr Wyld's** large model of the Earth. Great Globe, Leicester Square.
Advertisement for half-hourly demonstrations. Admission one shilling.
(3) $7\frac{1}{2}$ x 5

459 **Mr Wyld,** Leicester Square.
Colossal Globe. Large model of the Earth is now open.
(3) $3\frac{7}{8}$ x $4\frac{5}{8}$

460 **John Yarwell,** who hath lived many Years in St Paul's Church-Yard, is now removed to the Sign of Archimedes in Ludgate-street, the first Spectacle-Shop next Ludgate.
                                                     1697
Spectacles and optical instruments.
(1) $8\frac{3}{4}$ x $6\frac{3}{8}$
Plate 60

461 **John Yarwell,** at the Archimedes and 3 pair of Golden Spectacles in Ludgate-street, the Shop next Ludgate, London.
                                                     1697
Makes True Spectacles and other optical instruments.

The picture used at the top of item 460 is repeated here with the addition of three pairs of spectacles on the bridges of which appear respectively the three words 'the', 'Wright' & 'True'. The text is a redraft of that of item 460.
(1) $9\frac{5}{8}$ x $7\frac{7}{8}$

462 **John Yarwell,** at ye Archimedes and Spectacles in St Pauls Church yard, London. Optical Instrument Maker.   1683
(3) Photograph. $5\frac{1}{4}$ x 8. Two copies.

463 **John Yarwell** at the Archimedes and Three Golden Prospects, near the great North-Door in S. Paul's Church-Yard: London.
                                                     1692
Optical instrument maker.
(2) $4\frac{3}{4}$ x $6\frac{1}{8}$
Plate 57

464 **John Yarwell,** at the Archimedes and Three Golden Prospects, in St Paul's Church-Yard, London.   1694
Optical instrument maker.

Advertisement from 'The Present State of Europe', Vol. 5, Sept. 1694.
(2) $3\frac{1}{8}$ x $5\frac{1}{8}$

465 **John Yarwell** at the Archimedes and Three Golden Prospects, in St Paul's Church-Yard, London.   1694
Maker of True Spectacles and other optical instruments.

Advertisement from 'The Present State of Europe', Vol. 5, Nov. 1694.
(2) $2\frac{7}{8}$ x $4\frac{3}{4}$

Yarwell, *see* Timothy Brandreth and George Willdey, item 59.

466 **A. Yeates,** 12, Brighton Place, New Kent Road, London.
Mathematical Instrument Maker.
(1) $2\frac{3}{8}$ x $3\frac{1}{8}$

467 **Yeates & Son,** 2, Grafton Str^t, Dublin.
Optical, Mathematical & Philosophical Instrument Makers.

An adaptation of the trade cards used by Bleuler, item 52, and by Field, item 148.
(1) $2\frac{1}{4}$ x $3\frac{1}{2}$

468 **Yeates & Son,** Dublin.

Engraved picture of a measuring microscope removed from the instrument box containing it.
(3) $6\frac{3}{4}$ x $5\frac{1}{8}$

469 **H.R.H. the Duke of York,** 4, Thornhill Crescent, Barnbury, N.
Leather case manufacturer.

Manuscript bill with printed billhead, dated August 20th, 1876, for £1-3-0 for folder sheaths, etc., supplied to Messrs. Hudson & Son.
(3) $4\frac{1}{8}$ x $5\frac{1}{8}$

Reference numbers for catalogue items

Inventory Number   Almost all the items have been given a number in the Science Museum's inventory. Cards acquired together in a batch were often given a single inventory number which is common to all in the batch.

Location   Most of the items are contained in seven albums numbered I to VII, and the location is indicated by the album number and page. Mr. Court gave the Museum two albums of cards (numbers I & II) in which he had numbered the items serially. His numbers, preceded by C.C. (Court Collection), are given in the location column.
For those cards which are still pasted in numbered instrument boxes, the box number is given. The more substantial cards are glazed and framed, and as their location may vary it is not stated in the list.

Photographs   Where the Museum has an official photographic negative of any item, the negative number is given in the fourth column.

Catalogue No.	Science Museum Inventory No.	Location	Science Museum Negative No.
1	1934–121	C.C.75	881/70
2	1934–121	C.C.8	
3	1934–121	C.C.90	
4	1934–121	C.C.60	
5	1934–103		1005
6	1951–685	III 8	1112/52
7		VII 26	
8		VI 1	
9		VI 3	2495
10	1934–121	C.C.35	
11	1951–685	III 10	
12		VII 15	384/68
13	1934–121	C.C.13	
14	1934–121	C.C.28	
15	1951–685	III 14a	
16	1951–685	III 14b	
17	1934–122	C.C.159	274/69 1010
18	1934–99		830/67
19	1948–397	III 22	968/68
20	1948–397	III 16	
21	1934–101		501/68
22	1934–122	C.C.153	
23	1934–121	C.C.118	
24	1951–687	III 18	
25	1951–687	III 20	
26	1934–121	C.C.60	
27	1918–247		681/68
28	1934–116		836/67
29	1934–122	C.C.156	
30	1934–122	C.C.157	
31	1948–397	III 24	
32	1951–685	III 26a	
33	1951–685	III 28	
34	1951–687	III 30a	
35	1951–687	III 30b	883/70
36	1948–397	III 32	
37	1948–397	III 34	
38	1951–685	III 36	
39	1934–121	C.C.122	
40	1951–685	III 38	
41	1951–685	III 42	755/68
42	1951–685	III 44	
43	1934–122	C.C.172	275/69
44	1948–397	III 40	
45	1928–135	Box 803	714/68
46	1951–685	III 46	
47	1934–121	C.C.55	
48	1948–397	III 48	
49	1934–121	C.C.79	
50	1951–685	III 50	
51	1934–122	C.C.146	
52	1934–122	C.C.147	1013
53	1934–122	C.C.148	1014
54	1934–121	C.C.59	
55	1934–121	C.C.138	
56	1951–685	III 52	
57	1948–397	III 26b	
58	1934–121	C.C.17	
59	1951–687	III 54	
60	1951–687	III 56	884/70
61	1934–112		69/39
62	1951–685	III 58a	
63	1934–122	C.C.179	
64	1934–108		68/39
65	1934–121	C.C.77	
66	1948–397	III 58b	756/68
67	1951–687	III 60	
68	1934–121	C.C.84	

Catalogue No.	Science Museum Inventory No.	Location	Science Museum Negative No.
69	1934–121	C.C.25	
70	1951–687	III 62	
71	1934–124		601/50
72	1948–397	III 64a	
73	1934–121	C.C.97	
74	1948–397	III 64b	
75	1948–397	III 66	
76	1934–121	C.C.110	
77	1934–121	C.C.113	
78	1934–121	C.C.47	
79	1911–229	Box 382	690/68
80	1925–389	Box 730	694/68
81	1934–121	C.C.123	
82	1934–121	C.C.82	
83	1951–691		33/53
84	1934–121	C.C.71	
85	1934–121	C.C.73	
86	1934–121	C.C.23	
87	1951–687	III 68 III 69	
88	1951–685	III 70	
89	1948–397	III 72	
90	1934–121	C.C.128	
91	1951–685	III 74	757/68
92	1934–122	C.C.171	
93	1934–121	C.C.48	
94	1911–232	Box 389	708/68
95	1951–685	III 76	758/68
96	1934–109		833/67
97	1934–122	C.C.175	
98	1951–687	III 78	
99	1951–687	III 80	
100	1951–685	III 82a	
101	1934–119		1007
102	1934–122	C.C.167	
103	1951–690		
104	1934–122	C.C.177	
105		VII 32	
106	1934–115		70/39
,,	1951–685	III 88	
107	1934–121	C.C.106	
108	1934–121	C.C.52	
109	1951–685	III 86	
110	1951–685	III 84	
111	1934–121	C.C.15	
112	1951–685	III 82b	
113	1934–121	C.C.2	
114	1934–121	C.C.46	
115	1951–687	III 90	
116	1934–121	C.C.122	
117	1934–121	C.C.53	
118	1934–121	C.C.56	
119	1951–685	III 94	
120	1948–397	III 92	885/70
121	1951–685	III 96a	
122	1948–397	III 96b	
123		III 98	
124	1951–685	III 98	
125	1934–121	C.C.92	
126	1948–397	III 100	886/70
127	1948–397	III 102	
128	1934–121	C.C.78	
129	1934–121	C.C.83	
130	1951–685	III 104 VII 60	
131	1934–121	C.C.104	
132	1934–98		829/67
133	1951–685	III 110a & b	
134	1951–685	III 108c	
135	1951–685	III 108b III 110c	
136	1951–685	III 108c VI 11a	
137	1948–397	III 112b	
138	1951–685	III 112c	
139	1951–685	III 112a	
140	1951–685	VI 11b	
141	1951–687	III 114a	
142	1951–687	III 114b	
143	1934–121	C.C.91	
144	1934–121	C.C.6	
145	1914–888	Box 559	692/68
146	1948–397	III 118	
147	1951–685	III 116	
148	1951–685	III 120a	
149	1951–687	III 120b	
150	1934–121	C.C.114	
151	1934–121	C.C.29	
152	1934–121	C.C.10	
153	1951–685	III 122	
154	1948–397	IV 4	
155	1934–121	C.C.132	
156	1934–96		82/62
157	1948–397	IV 6	887/70
158	1951–685	IV 8	
159	1948–397	IV 10	
160	1948–397	IV 12a	
161	1948–397	IV 12b	
162	1948–397	IV 14	
163	1934–121	C.C.85	
164	1948–397	IV 18	
165	1951–685	IV 16	
166	1934–122	C.C.143a	
167		VI 47	
168	1934–122	C.C.154	
169	1951–685	IV 20	888/70
170	1951–685	IV 22	
171	1951–685	IV 26	
172	1951–685	IV 28	
173	1934–121	C.C.87	
174	1934–121	C.C.67	
175	1934–121	C.C.72	
176	1951–685	IV 30	889/70
177	1934–121	C.C.37	
178	1951–687	IV 32a & b	
179	1951–685	IV 32c	
180	1934–121	C.C.107	
181	1934–121	C.C.105	
182	1934–122	C.C.165	276/69
183	1948–397	IV 34a	
184	1934–121	C.C.65	
185	1934–122	C.C.160	
186	1951–685	IV 36	
,,	1934–122	C.C.173	
187	1951–687	IV 38	
188	1951–685	IV 40	
189	1934–121	C.C.19	
190	1948–397	IV 42	
191	1934–121	C.C.61	
192	1934–121	C.C. 30	
193	1889–68	Box 477	711/68
194	1934–102		1004
195	1934–121	C.C.98	
196	1951–685	IV 44	
197	1934–121	C.C.69	
198	1934–121	C.C.124	
199	1934–121	C.C.22	
200	1948–397	IV 46	890/70
201	1948–397	IV 48	
202	1948–397	IV 50	
203	1934–121	C.C.33	
204	1934–122	C.C.190	
205	1951–685	IV 52	
206	1929–443	Box 100	697/68

Catalogue No.	Science Museum Inventory No.	Location	Science Museum Negative No.
206	1929–444	Box 102	698/68
207	1948–397	IV 54	
208	1948–397	IV 58	
209	1934–110		1012
210	1948–397	IV 64	
211	1948–397	IV 66	
212	1934–121	C.C.141	
213	1905–125	IV 74a	
214	1934–121	C.C.38	
215	1934–121	C.C.36 / C.C.20	
216	1934–121	C.C.39	
217		IV 74b	
218	1934–121	C.C.1	
219	1951–367	Box 500	590/51
220	1951–685	IV 70	
221	1951–685	IV 72	
„	1934–121	C.C.4	
„	1915–312		6289
222	1934–121	C.C.40	
223	1951–687	III 106	
224	1948–397	IV 76	
225	1948–397	IV 78	891/70
226	1948–397	IV 80	
227		VI 21b	5960
228	1934–121	C.C.16	
229	1948–397	IV 86	
230	1934–121	C.C.134	
231	1951–687	IV 88	
232	1934–121	C.C.115	
233	1909–136	Box 323	689/68
234	1934–121	C.C.74	
235	1934–121	C.C.64	
236	1934–121	C.C.14	
237	1934–121	C.C.62	
238	1951–685	IV 90	
239	1934–100		1006
240	1951–687	IV 94	
241	1934–121	C.C.140	
242	1951–685	IV 96	
243		IV 82a	
244	1876–326	IV 82b	
245	1934–122	C.C.155	
246	1948–397	IV 128	
247	1934–121	C.C.18	
248	1934–121	C.C.26	
249	1951–685	IV 130	
250	1951–685	IV 106	
251	1934–97	828/67	
252	1951–685	IV 108	
253	1951–687	IV 110	
254	1934–121	C.C.109	
255	1951–685	IV 112	
256	1951–687	IV 114	892/70
257		VII 58	893/70
258	1951–685	IV 116	83/62
259	1934–122	C.C.178	
260	1951–687	IV 118	
261	1934–121	C.C.96	
262	1934–122	C.C.182	
263	1951–687	IV 120a	
264	1951–687	IV 120b	
265	1951–687	IV 122a	
266	1951–687	IV 122b	
267	1934–121	C.C.81	
268	1934–121	C.C.11	
269	1934–121	C.C.80	
270	1934–121	C.C.89	
271	1951–685	IV 126	
272	1951–685	IV 132	
273	1951–687	IV 134	
274	1934–122	C.C.158	
275	1934–122	C.C.188	
276	1948–397	IV 136	
277	1951–687	IV 138	
278	1948–397	IV 140	894/70
279	1948–397	IV 142	
280	1948–397	IV 144	
„	1951–685	VI 27	
281	1934–118		1022
282	1951–685	IV 150 / IV 152	
283	1934–121	C.C.76	
284	1934–122	C.C.150	
285	1948–397	IV 154	
286	1934–121	C.C.111	
287	1865–6	Box 262	688/68
288	1869–48	Box 277	706/68
289	1934–121	C.C.51	
290	1934–122	C.C.162	
291	1951–687	IV 158	
292	1934–121	C.C.88	
293	1934–122	C.C.144	
294	1951–685	IV 164 / IV 166	
295	1951–685	IV 162	
296	1948–397	IV 168	895/70
297	1948–397	IV 170	
298	1934–121	C.C.116	
299	1951–685	IV 172	
300	1911–226	Box 379	707/68
„	1876–992	Box 103	684/68
„	1876–993	Box 111	687/68
„	1911–231	Box 384	691/68
„	1911–244	Box 406	695/68
301	1948–397	IV 174	
302	1911–249	Box 410	696/68
303		VII 17	385/68
304	1934–121	C.C.143	
305	1951–685	IV 176b	
306	1951–687	IV 178	
307	1951–685	IV 180	896/70
308	1951–687	IV 182	
309	1934–117		1008
310	1934–121	C.C.119	
311	1934–122	C.C.152	1019
312	1951–687	V 4	
313	1948–397	V 6	
314	1923–523	Box 685	712/68
315	1934–122	C.C.149	
316	1951–685	V 8	
„	1934–95		71/39
317	1934–106		1011
318	1934–121	C.C.49	
319	1948–397	V 10	
320	1933–12	VI 49	
321	1934–121	C.C.58	
322	1934–121	C.C.57	
323	1934–121	C.C.142	
324	1951–687	V 12	
325		VI 53	
326	1934–121	C.C.129	
327	1934–121	C.C.112	
328	1934–121	C.C.54	
329	1934–121	C.C.94	
330	1934–121	C.C.3	
331	1934–121	C.C.120	
332	1934–121	C.C.121	
333	1951–685	V 14	
334	1934–111		834/67
335	1934–122	C.C.180	
336	1934–122	C.C.181	
337	1934–122	C.C.168	
338	1858–40	Box 808	716/68

Catalogue No.	Science Museum Inventory No.	Location	Science Museum Negative No.
339	1934–120		1002
340	1934–105		67/39
341	1934–122	C.C.183	
342	1934–121	C.C.130	
343	1951–685	V 18	
344	1951–685	V 20	
,,	1934–122	C.C.164	
,,	1948–397	VI 35	
345	1951–685	V 22	
346	1951–687	V 21	
347	1948–397	V 28	
348	1934–121	C.C.133	
349	1934–121	C.C.99	
350	1934–121	C.C.24	
351	1934–121	C.C.86	
352	1934–121	C.C.126	
353	1951–685	V 24 / V 26	
354		VII 19	
355	1910–188	V 30a	
356	1948–397	V 30b	
357	1934–122	C.C.170	
358	1951–687	V 32	
359	1934–114		84/62
360	1934–122	C.C.163	
,,	1951–685	V 34	
361	1934–121	C.C.117	
362	1951–685	V 36	
363	1948–397	V 38	897/70
364	1951–685	V 40	
365	1934–121	C.C.34	
366	1948–397	V 42	
367	1948–397	V 44	
368	1951–685	V 46	
,,	1948–397	VI 37	
369	1951–685	V 52	
370	1951–685	V 50	
371	1951–685	V 48	
372	1951–685	V 54	
373	1951–685	V 56	
374	1951–685	V 58	
375	1948–397	V 60 / VI 39	
376	1934–121	C.C.31	
377		V 61	
378	1951–685	V 62	
379	1948–397	V 64	
380	1948–397	III 12	882/70

Catalogue No.	Science Museum Inventory No.	Location	Science Museum Negative No.
381	1934–104		831/67
,,	1951–686		
382	1951–685	V 66	
383	1948–397	V 68	
384	1948–397	V 70a	
385	1948–397	V 70b	
386	1934–122	C.C.186	
387	1951–685	V 72	
388	1934–122	C.C.187	93 is similar
389	1951–685	V 74a	
390	1951–685	V 74b	
391	1934–121	C.C.95	752/68
392	1934–121	C.C.93	
393	1934–121	C.C.50	
394	1934–121	C.C.70	
395	1934–122	C.C.184	753/68
396	1934–122	C.C.185	
397	1948–397	V 76	
398	1948–397	V 78	
399	1934–121	C.C.63	
400	1951–687	V 80	
401	1951–687	V 82	
402	1918–122	Box 789	713/68
403	1933–11	V 76	
404	1931–95	Box 917	693/68
405	1929–936	Box 893	715/68
406	1915–130	V 83	
407	1934–113		835/67
408	1948–397	V 86	
,,	1934–123		1003
409	1951–687	V 84	
410	1934–121	C.C.100	1023
411	1934–121	C.C.101	
412	1934–121	C.C.102	
413	1948–397	V 88	
414	1934–121	C.C.44	
415	1934–121	C.C.139	
416	1934–121	C.C.42	
417	1934–121	C.C.43 / C.C.45	
418	1934–121	C.C.108	
419	1934–121	C.C.41	
,,	1951–687	V 90	
420	1951–687	V 92	
421	1948–397	V 94 / V 96	
422	1948–397	V 98b	

Catalogue No.	Science Museum Inventory No.	Location	Science Museum Negative No.
423	1951–685	V 98a	
424	1934–121	C.C.136	791/68
425	1876–998	Box 78	682/68
426	1934–121	C.C.27	
427	1934–122	C.C.161	1018
428	1934–122	C.C.151	
429	1934–121	C.C.68	
430	1934–122	C.C.145	
431	1934–107		832/67
432	1948–397	V 100	
433	1948–397	V 102	
434	1951–685	V 104	
,,	1934–122	C.C.166	
435	1951–687	V 110	
436	1951–687	V 108a	
437	1948–397	V 106	
438	1951–687	V 108b	
439	1934–121	C.C.137	
440	1951–687	V 108c	
441	1934–121	C.C.9	
442	1934–122	C.C.189	
443	1948–397	V 112	
444	1934–94		1009
445	1951–689		
446	1951–687	V 114	
447	1951–685	V 116	
448	1951–687	V 118	
449	1948–397	V 120	
450	1934–121	C.C.7	
451	1934–121	C.C.21	
452	1951–685	V 122	
453	1951–685	V 124a	
454	1951–687	V 124b	
455	1951–685	V 126	754/68
456	1951/685	V 128	
457		VI 53	
458	1951–687	V 130	
459	1934–121	C.C.103	
460	1948–399		523/50
461	1951–685	V 134	898/70
462	1951–687	V 135	
,,	1951–688		
463	1934–121	C.C.125	1025
464	1934–121	C.C.127	1024
465	1934–121	C.C.131	
466	1934–121	C.C.12	
467	1951–685	V 136	
468	1892–13	V 140	
469	1951–685	V 138	

Index
The numbers are the numbers of the items in the catalogue.

The index is only of the words used in the catalogue. There are many more instruments mentioned on the actual trade cards themselves (eg Plates 24, 32 & 38), which have had to be omitted for economy of space. Nor have all the words used in the catalogue been indexed. For example, many of the advertisers describe themselves as makers of Mathematical, Optical and Philosophical Instruments. Because there are so many it has not been thought useful to mention them individually in the index. Opticians, which included those who made optical instruments as well as spectacle-makers, are also too numerous to be usefully itemised in the index. Mathematical instruments, especially, are frequently (157 times) mentioned on the cards. In the early days the description 'Mathematical Instruments' covered a large part of what we should now call 'Scientific instruments'. Later, optical and philosophical instruments were distinguished from the mathematical ones. Then gradually the word philosophical went out of fashion and the words scientific and physical began to be used. Those later cards using 'astronomical', 'physical' or 'scientific' in their descriptions have been mentioned in the index as a matter of interest.

Index of occupations and wares

Artist, Engraver   46, 47, 170, 194, 229, 331/2, 356, 376, 417/8

Astronomical instruments   130, 193, 213–6, 287/8, 411, 435

Axeltrees   159

Barometers, Weather-glasses   30, 36, 46, 113, 185, 294/5, 305, 321/2, 374, 385

Binnacle   391

Blowing & pumping machines   67, 88, 115, 146, 181, 261

Book-binder   66

Book-seller, Stationer   10, 66, 99, 111, 150, 184/5, 187, 220, 235, 284, 289, 292, 307, 320, 337, 348, 393, 423, 447

Brass- or Iron-founder   211, 272

Bug destroyer   98

Camera obscura   131

Cards *see* Mathematical Cards

Case-maker   15, 169, 469

Charts *see* Maps

Chemical apparatus   197/8, 414, 419, 429, 452

Chemist   11, 40, 121/2, 198, 207, 280, 432, 437

Chirurgical (Surgical) instruments   115, 447

Chronometers *see* Clockmaker

Clock, watch & chronometer maker   16, 49, 69, 151, 226, 247/8, 267, 375, 382, 394, 398, 433, 441

Clockmakers' tools   31, 164

Compasses, magnetic *see* Magnets

Cutlery   112

Dentist   333

Dials & dialling instruments   106, 150, 170, 184, 272, 308, 346, 386/7, 409, 412

Drawing instruments   325, 414, 418

Druggist *see* Pharmacist

Electrical & Electro-magnetic apparatus   84, 172

Engineer   86–8, 159, 319, 362

Engraver, *see* Artist

Engraver's glasses   363

Exhibitions, Museums   12, 71, 74, 298, 303, 328/9, 422, 435, 458/9

Experimental philosophy, instruments
    for   115, 158
Flags, Bunting   58, 325, 371
Foundry *see* Brass-founder
Fortification, instruments for   409
Gauging instruments   77, 134/5, 139,
    150, 241, 409
Geodetical instruments   45
Gilder   66
Glass-making & -blowing   12, 20,
    303, 340, 424
Globes   110, 179, 184/5, 192, 194/5,
    220, 235, 285–8, 342, 347–53, 356,
    410, 458/9
Haberdasher   354
Hydrometers   34/5, 62, 133–5, 139,
    154, 241, 302, 310
Hygrometers   30
Iron-founder *see* Brass-founder
Jeweller   151, 226, 394, 445
Kaleidoscopes   223
Lanterns *see* Projection lanterns
Lamps   211, 397
Levels   186
Magic lanterns *see* Projection lanterns
Magnets, Magnetic compasses   8, 31,
    49, 64, 89, 97, 123, 136, 203, 227,
    236/7, 262, 372
Maps, Charts   14, 32, 34/5, 110/1,
    136, 160/1, 184, 189, 192, 235, 262,
    275, 297, 302, 320, 342, 347-51,
    353, 356, 376, 393, 423
Matches   207, 432
Mathematical cards   409
Mathematical instruments. Many items
Meteorological instruments 369
Microscopes   71, 101, 103, 109, 114,
    131, 162, 254, 261, 304, 306/7, 318,
    323/4, 414, 448, 468
Mineralogical instruments   319
Mineralogist   121
Mirrors   305
Museums *see* Exhibitions
Music   291, 421, 447
Nautical Almanack   43
Nautical instruments   32, 43, 58, 64,
    94, 108, 111, 113, 136, 159, 189,
    196, 203, 233, 236/7, 247/8, 292/3,
    299, 320, 325, 334, 337, 344/5, 369–
    74, 378, 380, 389–92, 409, 423, 451
Oculist   165, 190, 313, 364, 436
Opera glasses   112, 127
Optical instruments. Many items

Optician. Many items
Orreries, Planetariums   90–2, 220,
    285, 298, 383, 422
Pedometers   202, 375
Pelorus   374
Pharmacist   11, 122, 242, 280, 400/1
Philosophical instruments. Many items
Photographer   39, 81, 114, 199,
    331/2
Photographic apparatus   198, 271,
    452
Physical instruments   65, 171, 319
Planetariums *see* Orreries
Polarising apparatus   47
Projection lanterns   47, 74/5, 141
Presses   159
Pumping machines *see* Blowing
    machines
Quadrants   136, 203, 370, 388, 390,
    392
Quintants   168
Saccharometers   34/5, 134/5, 137,
    139, 241
Scale- (ie Balance-) maker   33, 37, 61,
    93, 152, 191, 274, 283, 413
Scales *see* Slide-rules
Scientific instruments   78/9, 87, 139,
    243/4
Sectors   63, 187, 395/6
Sextants   58, 136, 189, 203, 370
Ship-models   205
Shipping, Ships' supplies *see* Nautical
    instruments
Ships Chandler   64, 293, 378
Ships logs   159
Slide-rules, Carpenters' rules, Linear
    scales   77, 184, 204, 327, 335/6,
    341, 387, 393, 442
Spectacle-maker   10, 17, 21, 23/4, 59,
    101, 112, 120, 128, 157, 174, 200/1,
    208–10, 224/5, 255, 265, 273,
    278/9, 296, 301, 307, 340, 366, 380,
    443, 451, 460/1, 465
Spectroscopes   374
Stationer *see* Bookseller
Stereoscopes   420
Sundials *see* Dials
Surgical instruments *see* Chirurgical
    instruments
Surveying instruments   150, 229, 314,
    369, 372–4, 409
Surveyor   110, 155, 230, 312, 315
Teacher   99, 111, 147, 173, 180, 228,
    232, 260, 289

Telescopes   25, 58, 104, 112, 119,
    131, 136, 153, 177, 203, 212, 217,
    240, 259, 261, 263, 265, 282, 321/2,
    368/9, 371, 373/4, 393, 414/5, 435
Theodolites   186
Thermometers   30, 36, 46, 113, 139,
    294, 305, 321/2
Toys   36, 163, 445/6
Watch-maker *see* Clock-maker
Watch-springer   169
Waywiser *see* Pedometer
Weather-glasses *see* Barometers
Weapons   50

Geographical index

AMERICA
    San Francisco, Cal.   321, 322
AUSTRIA
    Vienna   14
FRANCE   231, 235
    Brest   94
    Marseille   325
    Paris   16, 40, 48, 50, 85–7, 93,
        141/2, 160/1, 171, 190, 206, 229,
        234, 243/4, 267, 290, 297, 318/9.
        356, 420
    St. Malo   94
GERMANY
    Berlin   425
GREAT BRITAIN
    Bath   2
    Berwick-on-Tweed   393
    Birmingham   37, 72, 223
    Brighton   128/9, 331/2
    Bristol   34, 49, 223, 435, 454
    Cambridge   121, 261, 304
    Cardiff   49
    Cheltenham   2, 3, 116, 280
    Chester   394
    Chichester   260
    Clerkenwell (London)   15, 383, 398
    Dalkeith   165
    Deptford (London)   433
    Devonport   34
    Edinburgh   143, 223, 272
    Faversham   11
    Glasgow   62, 247/8
    Greenock   247
    Greenwich (London)   202, 226
    Horslydown   76, 205
    Hull   303

GREAT BRITAIN: cont.

  Ipswich   224
  Islington (London)   382
  Leamington   122
  Leeds   397
  Leith   165
  Limehouse (London)   236, 320
  Liverpool   1, 34, 69, 82/3, 116, 154, 196, 223, 389/90, 423, 447
  London   Many items
  Manchester   4, 39, 114, 151
  Musselburgh   165
  Newcastle   10, 68, 163, 270/1, 305, 366, 451
  Newington (London)   400
  Norwich   224, 354
  Oxford   261, 308
  Portsea   233
  Portsmouth   34
  Reading   261
  Rotherhithe (London)   77, 337, 391/2
  St. Day (Cornwall)   449
  Sarum   261
  Sheffield   44, 82–4, 112/3, 153
  South (Lincolnshire)   147
  Southampton   379/80
  Southwark (London)   67
  Southsea   233
  Walworth (London, Surrey)   46, 400/1
  Wapping (London)   64, 111, 180, 346, 378
  Windsor   75
  Woolton (Lancashire)   31
  York   170

IRELAND
  Belfast   34
  Cork   34
  Dublin   34, 245/6, 268/9, 467/8

ITALY
  Genoa   70
  Milan   57

NETHERLANDS
  Antwerp   276

SPAIN
  Gibraltar   299
  Madrid   273, 314

SWEDEN
  Stockholm   45

SWITZERLAND
  Geneva   16

Languages other than English

Dutch: 339
French: 14, 25, 48, 50, 70, 85–7, 93/4, 141/2, 160/1, 171, 190, 206, 229, 231, 234/5, 240, 243/4, 255, 277, 290, 297, 318/9, 325, 339, 356, 407, 420
German: 255, 425
Italian: 57, 70
Spanish: 25, 273, 299, 314

The Plates

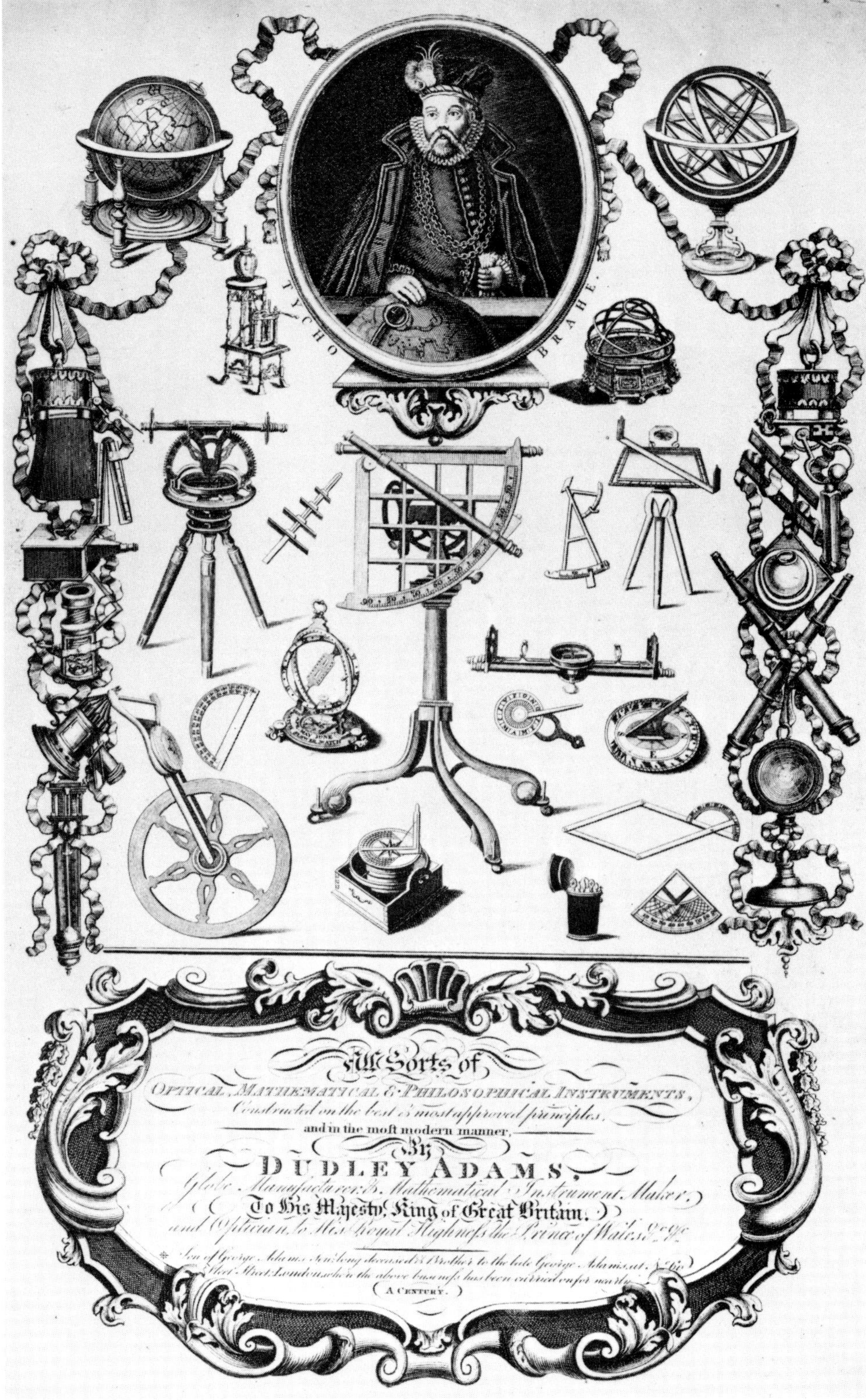

1   Dudley Adams
catalogue no.5

GEORGE ADAMS,

MATHEMATICAL INSTRUMENT-MAKER TO HIS MAJESTY,

At TYCHO BRAHE's HEAD, in FLEET-STREET, LONDON,

MAKES and SELLS all Sorts of the moſt curious MATHEMATICAL, PHILOSOPHICAL, and OPTICAL INSTRUMENTS, in Silver, Braſs, Ivory, or Wood, with the utmoſt Accuracy and Exactneſs, according to the lateſt and beſt Diſcoveries of the modern MATHEMATICIANS.

HADLEY's QUADRANTS, with the lateſt Improvements, in the moſt exact Method, with Glaſſes whoſe Planes are truly parallel.

Azimuth and Steering Compaſſes, invented by Dr. *Gowin Knight*, F. R. S. approved of, and uſed by his Majeſty's Royal Navy.

N. B. Theſe Compaſſes are all examined and certified by Dr. *Knight*.

Large ASTRONOMICAL QUADRANTS, TRANSIT and EQUAL ALTITUDE INSTRUMENTS, for obſerving the Tranſits of the Sun and Stars over the Meridian, &c.

Sun Dials HORIZONTAL, for Pedeſtals in any Latitude; with Variety of PORTABLE ones, either UNIVERSAL, or for ſeveral different Latitudes, with new Improvements.

Choice of curious Caſes of DRAWING INSTRUMENTS, in Silver, Braſs, &c. containing a Sector, Scales, Proportionable, and other Compaſſes; Drawing-Pens, a Protractor, Parallel Rules, &c.

A new-invented Portable MICROSCOPE for viewing all Kind of Minute Objects, as well Opake as Tranſparent, in ſo conſpicuous and conciſe a Manner, as to comprehend all the Uſes of all the other Sorts of Microſcopes in one Apparatus; and magnifies to ſo great a Degree, as to diſcover the Circulation of Blood in Animals, the Periſtaltic Motion of Inſects, the Farinæ of Vegetables, and many other ſurpriſing Phænomena, otherwiſe not perceptible.

The double Conſtructed MICROSCOPE; Mr. *Ellis*'s ÆQUATIC MICROSCOPE; SOLAR MICROSCOPES; MAGELLESCOPES, &c.

The New ACROMATIC TELLESCOPE, with a compound Object Glaſs, approved by all the Curious in Optics; with all other Sorts of REFRACTING TELLESCOPES; NIGHT TELLESCOPES, &c.

REFLECTING TELLESCOPES, of the lateſt Improvement.

MICROMETERS of the neweſt Conſtruction, elegantly fitted to Refracting or Reflecting Tellescopes.

ORRERIES and PLANETARIUMS, greatly improved.

Inſtruments proper for GUNNERY, FORTIFICATION, &c.

PANTOGRAPHERS, for reducing Drawings and Pictures of any Size, in the moſt complete Manner.

Inſtruments for taking the true Perſpective of any Landſcape, Building, Gardens, &c. and others for copying of Drawings.

New GLOBES; mounted in a peculiar Manner, whereby the Phænomena of the Sun, Earth, Moon, &c. are exhibited according to Nature.

AIR-PUMPS, or Engines, either for exhauſting or condenſing the Air, and this by turning one Cock only, with all their Appurtenances; whereby the Properties of that moſt uſeful Fluid are diſcovered and demonſtrated by undeniable Experiments; HYDROSTATICAL BALANCES, nicely adjuſted for determining the ſpecific Gravity of Fluids and Solids, &c.

Curious BAROMETERS, Diagonal, Wheel, Standard, or Portable, with or without Thermometers. Alſo the ſo much famed QUICKSILVER THERMOMETERS, made after any of the Forms.

THEODOLITES, of the lateſt Conſtruction; WATER LEVELS, which may be adjuſted at one Station; MEASURING WHEELS; Pocket and Coach WAY WIZERS, for meaſuring the Way, &c.

SPECTACLES ground on Braſs Tools, in the Manner approved of by the Royal Society, ſet in Variety of convenient Frames: Alſo READING GLASSES of all Sorts, ſet in Silver or other Metal, to turn into Caſes of various Kinds.

PRISMS, for demonſtrating the Theory of Light and Colours.

The CAMERA OBSCURA for drawing in Perſpective, in which all external Objects are repreſented in their proper Colours and exact Proportions.

CONCAVE, CONVEX, and CYLINDRICAL MIRRORS, OPERA GLASSES, MULTIPLYING GLASSES, SPECTACLES of the true *Venetian* Green Glaſs, MAGIC LANTHORNS, &c.

ZOGRASCOPES, for viewing Perſpective Prints.

N. B. Gentlemen may have any Model or Inſtrument made in Metal or Wood, with Expedition and Accuracy, and carefully packed up to be ſent to any Part of the World.

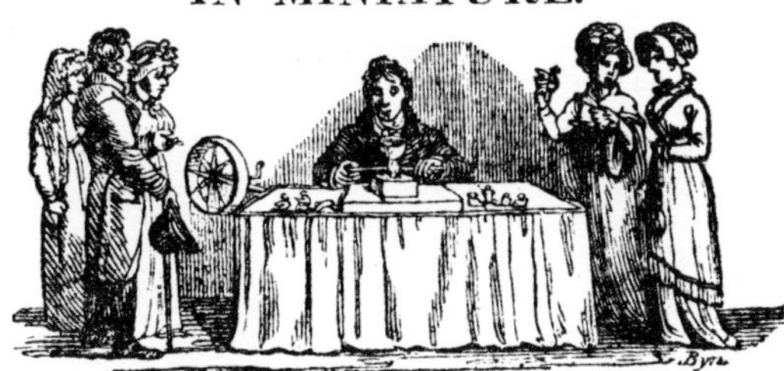

NEW EXHIBITION
OF
FANCY GLASS WORKING
IN MINIATURE.

Glass Spinning, Blowing, Variegating, Twisting, and Linking;
Is now Exhibiting in a great Variety of Articles,

At No. 4, BURLINGTON ARCADE, Piccadilly,
And at No. 24, ANDERSON'S BUILDINGS, CITY ROAD,
BY J. ANDREWS & SONS,
ARTISTS IN GLASS.

Open from 11 in the Morning till 8 at Night.----Admittance ONE SHILLING,
For which every Person may receive an Article to the Amount of their Admission.

HUNDREDS OF SPECIMENS MAY BE SEEN.

Messrs. A. make various kind of Devices, in all colours of Glass, before the Company, for sale ; and form Bird, and Quadrupeds while the Glass is in a fluid state, without the use of any kind of Instrument whatever. Also make Necklaces, Baskets of flowers, Ornamental Pens, Ships, Anchors, Crosses, Rings, Cupids, Doves, Witches, Mermaids, Swans, Horses, Dogs, Stags, &c. &c. Messrs A. will spin out of One Pound of common Window Glass, 2,000,000 Yards, at

ONE THOUSAND YARDS IN ONE MINUTE.

To be seen a complete Stag Hunt, enclosed in a common Nut-shell, 52 hounds, a stag and 8 horsemen. Also a Horse and Gig, in a Nut-shell, with many curious, entertaining, and droll subjects in Glass, viz. John Bull, Buonaparte, and his satanic Majesty at marbles; a very beautiful Groupe of old and young Witches in full sally; a Donkey race, highly finished ; a fine Brood of young Belzebubs, and the old Ones feeding them with worms; Happy David ; mother Shipton ; Devil and Dr Faustus at Wine (capital Likenesses,) Adam and Eve in Paradise, with birds, beasts, &c. Faith, Hope and Charity ; Lucky Tom; the hunted Dandy on his Hobby-horse; Billy Button bewitched and John Bull bedevilled ; the Ghostly Gambol ; Old Satan caught in a Steel Trap; the Triumph of Bacchus ; ditto of Venus ; ditto of Apollo ; Flora and Juno ; the Muses' Revel ; Alderman Guzzle's last Drop; Doctor Bolus fighting with Death ; John and Molly kissing under the Misletoe Bush ; Doctor Syntax ; Duke and Duchess of Mutton, and King of Pork ; John Doe and Richard Roe ; Flora's Bower ; Temple of the Muses ; the Honey Moon ; Solomon's Temple ; the Prince Regent's Pavilion ; Shepherd and Flock ; Off to Gretna Green; Old Taffy in Jack Boots, and a New Wig ; Don Quixote, Rozinante, his Dulcinea del Toboso and Sancho Panza in full trim ; a Nest of Young Cupids ; a Flock of Mermaids ; Europa on the Bull ; a Dance of Demons; a Flight of Spectres ; Britannia awake ; Tam o'Shanter's narrow Escape ; King Charles in the Oak ! Moses found in the Bullrushes ; the Dance of Death ; the contents of Noah's Ark ; the Spectres from the Tombs ; Don Juan, Johnny Gilpin's Exploits ; Devil take Love ; Prince Cobweb ; the Lunarian Travelling with his Head under his Arm ; with many more Flights of Fancy too numerous to insert in the limits of a Bill.

Proper Objects for the Kaleidescopes, at One Shilling per Dozen.

Glass Shades of all Sizes, to cover Figures, Vases, Shells, &c. made of the finest White Crystal Glass. Messrs. ANDREWS respectfully wish that this Bill may be presented to the Family of the House.

May the 1st 1820.

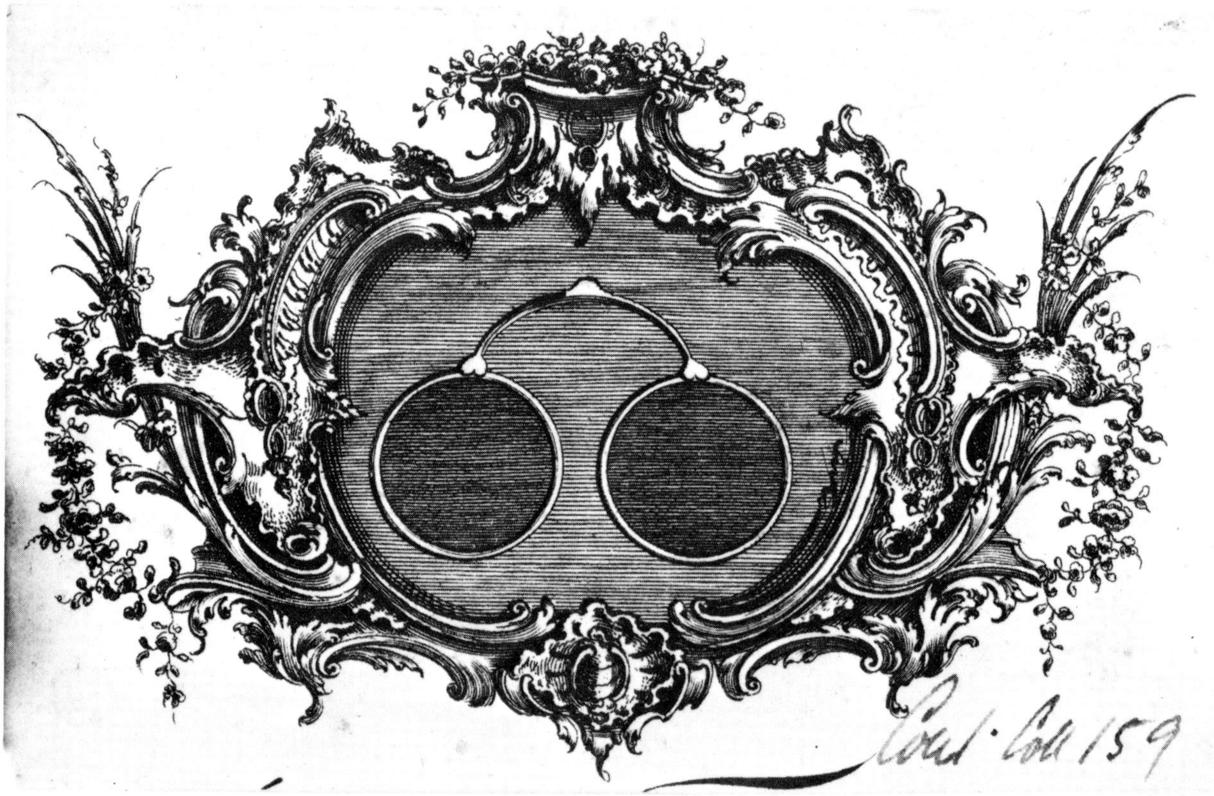

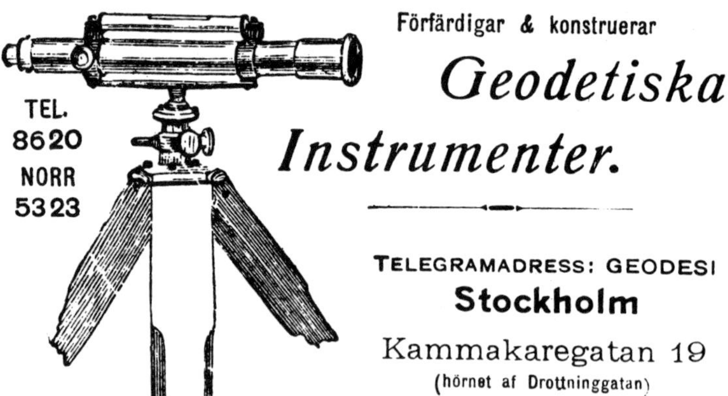

4 Attributed to
Ayscough
catalogue no.17

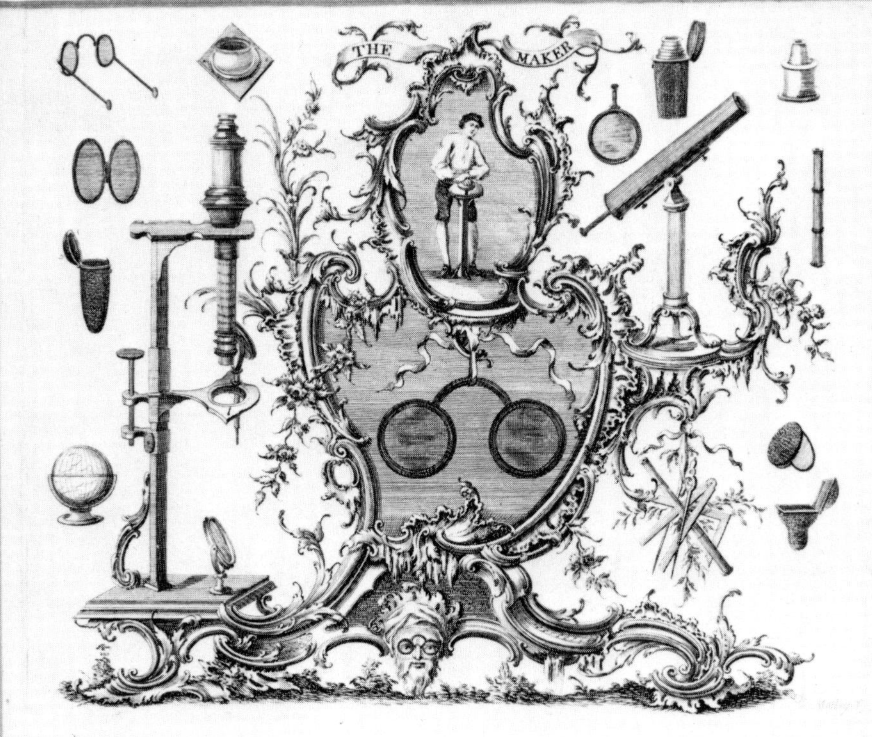

5 Fr. J. Berg
catalogue no.45

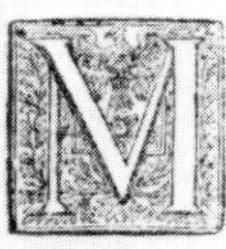

6 James Ayscough
catalogue no.18

ƎAMES AYSCOUGH,

OPTICIAN,

At the Great GOLDEN SPECTACLES, in *Ludgate-Street*,
near St. PAUL's, *LONDON*,

(*Removed from* Sir ISAAC NEWTON's HEAD *in the same Street*)

MAKES and SELLS, (Wholesale and Retail) SPECTACLES
and READING-GLASSES, either of *Brazil*-Pebbles, White, Green, or
Blue Glass, ground after the truest Method, set in neat and commodious Frames.

CONCAVES for SHORT-SIGHTED PERSONS.

REFLECTING and REFRACTING TELESCOPES of various Lengths, (some of
which are peculiarly adapted to use at Sea;) Double and Single MICROSCOPES, with
the latest Improvements; PRISMS; CAMERA OBSCURA's; Concave and Convex
SPECULUMS; MAGICK LANTHORNS; OPERA GLASSES; BAROMETERS
and THERMOMETERS; SPEAKING and HEARING-TRUMPETS; with all
other Sorts of Optical, as well as Mathematical and Philosophical Instruments.

Together with Variety of MAPS, and GLOBES of all Sizes.

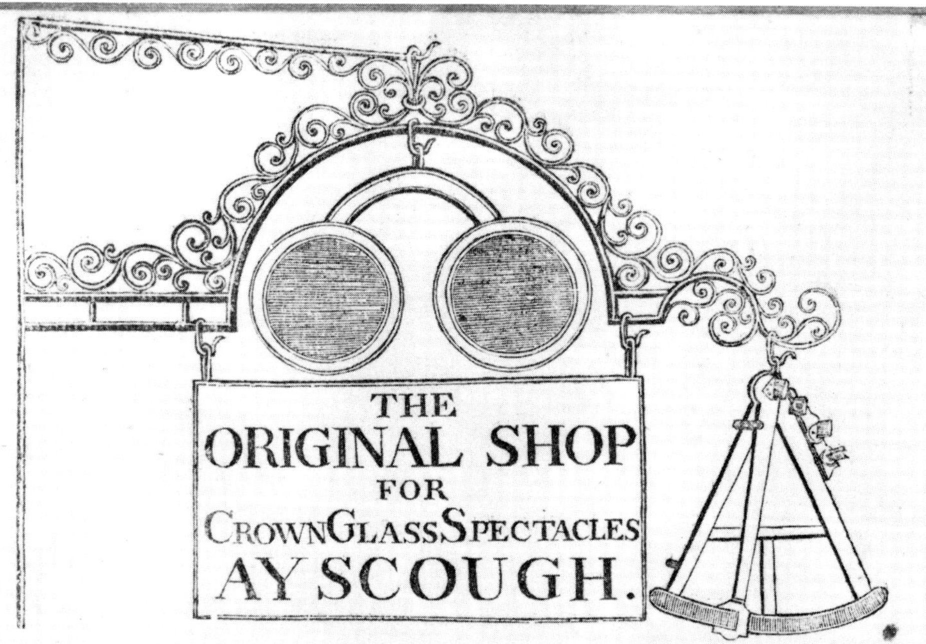

JAMES AYSCOUGH,
OPTICIAN,

At the Great GOLDEN SPECTACLES and QUADRANT,
N°. 33, in *Ludgate-Street*, near St. PAUL's, *LONDON*,

(The Original Shop for superfine CROWN-GLASS SPECTACLES.*)*

MAKES and Sells (Wholesale and Retail) SPECTACLES and READING-GLASSES, either of *Brazil* Pebbles, Crown, White, Green, or Blue Glass, ground after the truest Method, set in neat and commodious Frames, some of which are so contrived as to press neither upon the Nose nor the Temples.

CONCAVES for Short-sighted Persons.

MICROSCOPES of various Kinds, particularly one of a new Construction, which is used either as a single one or a double one, for transparent or opake Objects; the Magnifiers of which are adapted for viewing the smallest Object in Nature, as also those of the Size of a common Flower. It may be used with the Solar Apparatus.

REFLECTING TELESCOPES of all Sizes.

REFRACTING TELESCOPES of various Lengths, some of which are peculiarly adapted to use at Sea, and are so contrived to be adjusted in such a Manner as to fit any Eye, whether short-sighted, or the reverse. To which also is added a Conveniency for receiving or excluding more or less Light, according as the Brightness or Cloudiness of the Atmosphere requires. Which by several late Trials at Sea were found very advantageous. Also, a New-invented Achromatic TELESCOPE, which magnify much more, than those made on the common Principle.

PRISMS for demonstrating the Theory of Light and Colours. Scioptic BALLS.

CAMERA OBSCURA's for delineating Landskips and Prospects, Concave and Convex SPECULUMS, MAGICK-LANTHORNS, OPERA-GLASSES, &c.

OPTICAL MACHINES for viewing Perspective Prints.

BAROMETERS, Diagonal, Standard, or Portable.

THERMOMETERS, whose Scales are adjusted to the Bores of their respective Tubes; Hygrometers, and Hydrometers; Hydrostatical Balances; *Hadley*'s Quadrants; Globes; Cases of Drawing Instruments; Scales; Parallel Rulers; Sun-Dials; Rules; Pencils; with all other Sorts of Optical, Mathematical, or Philosophical Instruments.

THOMAS BARNETT,

(No. 61,)

GREAT TOWER-STREET, *LONDON*,

Optical, Mathematical and Philosophical Instrument-Maker to

HIS MAJESTY's

BOARDS of CUSTOMS and EXCISE,

Makes the following Articles, with a Variety of others, which are sold, Wholesale and Retail, at the lowest Prices.

SPECTACLES of all Sorts, mounted in Silver, Tortoiseshell, or Steel, with Glass or Brazil Pebble, adapted to the various Defects of Sight

Reading Glasses, set in Silver, Pearl, Tortoiseshell & Horn

Single, Double and Treble Magnifiers, set as above

Concave Glasses for short-sighted Persons

Reflecting and Refracting Telescopes for celestial and terrestrial Objects, with different magnifying Powers

Micrometers with the latest Improvements

Opera Glasses and Perspectives on the most improved Constructions

Solar Microscopes for transparent and opaque Objects

Single and Compound Microscopes

Aquatic, Botanic and other Microscopes, convenient for the Pocket

Magic Lanterns and Sliders to ditto

Camera Obscuras for delineating Landscapes

Optical Machines for viewing Perspective Prints

Concave and Convex Mirrors

Ditto for the Pocket, in black Fishskin and Morocco Cases

Black Convex Mirrors for the Purposes of Drawing

Cylinders and Cylindrical Mirrors

Prisms, for demonstrating the Theory of Light and Colours

Theodolites, Circumferentors, Plane Tables, Level Telescopes and other Surveying Instruments

Cases of Instruments fitted up for all Branches of the Mathematics

Circular Protractors

Sectors, Scales and Parallel Rules, Gauging Rules, Sliding Rules, and all other Sorts of Rules, of Silver, Brass, Ivory and Box-Wood

Triangle, Elliptical and Beam Compasses

Improved Gunter's Scales

Sliding ditto

Pentagraphs of Brass and Wood, for reducing Profiles, Maps, &c.

Sextants and Hadley's Quadrants

Steering and Cabin Compasses

Compasses for Miners, and for the Pocket

Loadstone and Steel Magnets

Gunner's Quadrants

Ditto Perpendiculars and Callipers

Orreries, Planetariums, and Tellurians, of various Sorts

Globes of all Sizes

Maps, Sea-Charts, and Navigation Books

Air-pumps, with different Apparatus

Condensing-Engines, Air-Guns, and Air-Fountains

Hydrostatic Balances carefully adjusted

Hydrometers of all Sorts, and Glass Bubbles, for ascertaining the different Strengths of Spirits, Worts, Acids and Alkalies

Glass Pumps and other Hydraulic Machines

Barometers for indicating the Approach of a Storm at Sea, and other Barometers, either Diagonal, Wheel, Standard, or Portable

Thermometers for Brewers, Distillers, Brandy-Merchants, Physicians, Hot-Houses, and other Purposes

Electrical Machines, Batteries and Electrometers, with a Variety of other Apparatus

Conductors for the Safety of Ships in Storms of Lightning

Ditto for the Preservation of Buildings, affixed with particular Care in Town or Country

Long Callipers, Cross Callipers, Bung Rods, Head Rods and Jerquers Rules, for Port Officers, &c.

Also Gauging and other Instruments, for Officers of Excise or Customs, in complete Sets, or any single Instrument

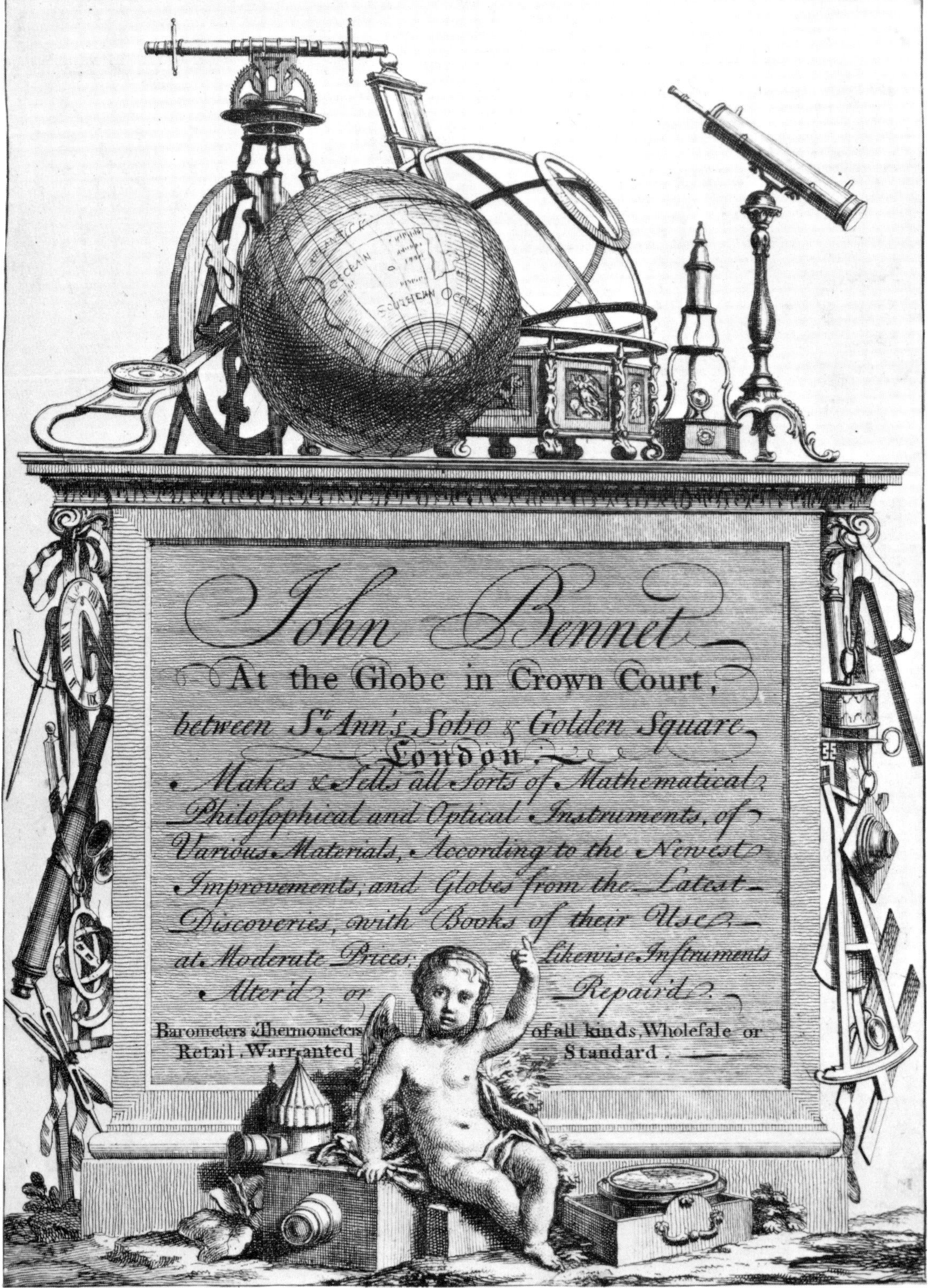

John Bennet
At the Globe in Crown Court,
between St. Ann's Soho & Golden Square
London.
Makes & Sells all Sorts of Mathematical,
Philosophical and Optical Instruments, of
Various Materials, According to the Newest
Improvements, and Globes from the Latest
Discoveries, with Books of their Use
at Moderate Prices,   Likewise Instruments
Alter'd, or   Repair'd.
Barometers & Thermometers   of all kinds, Wholesale or
Retail, Warranted   Standard.

left
9 John Bennett
catalogue no.41

right
10 John Bennett
catalogue no.43

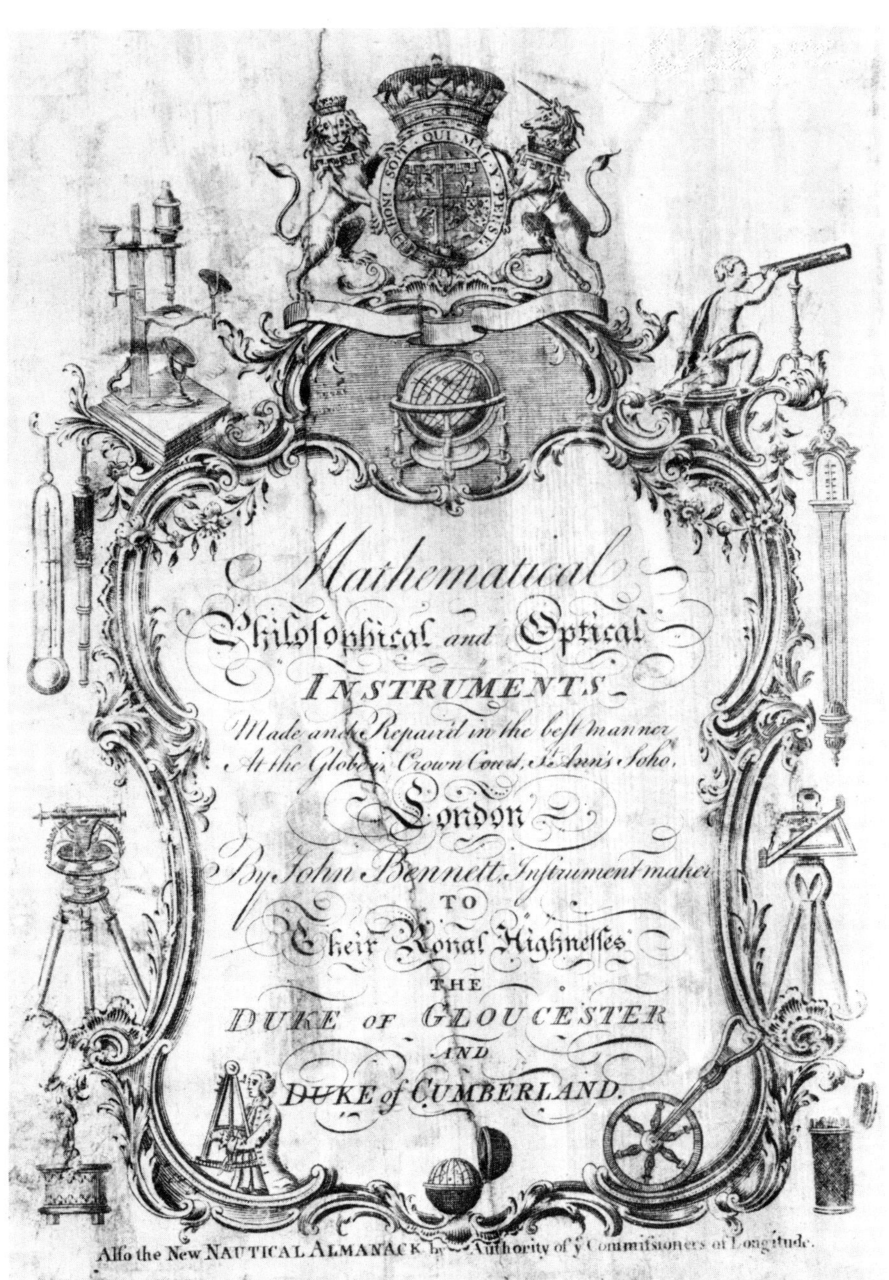

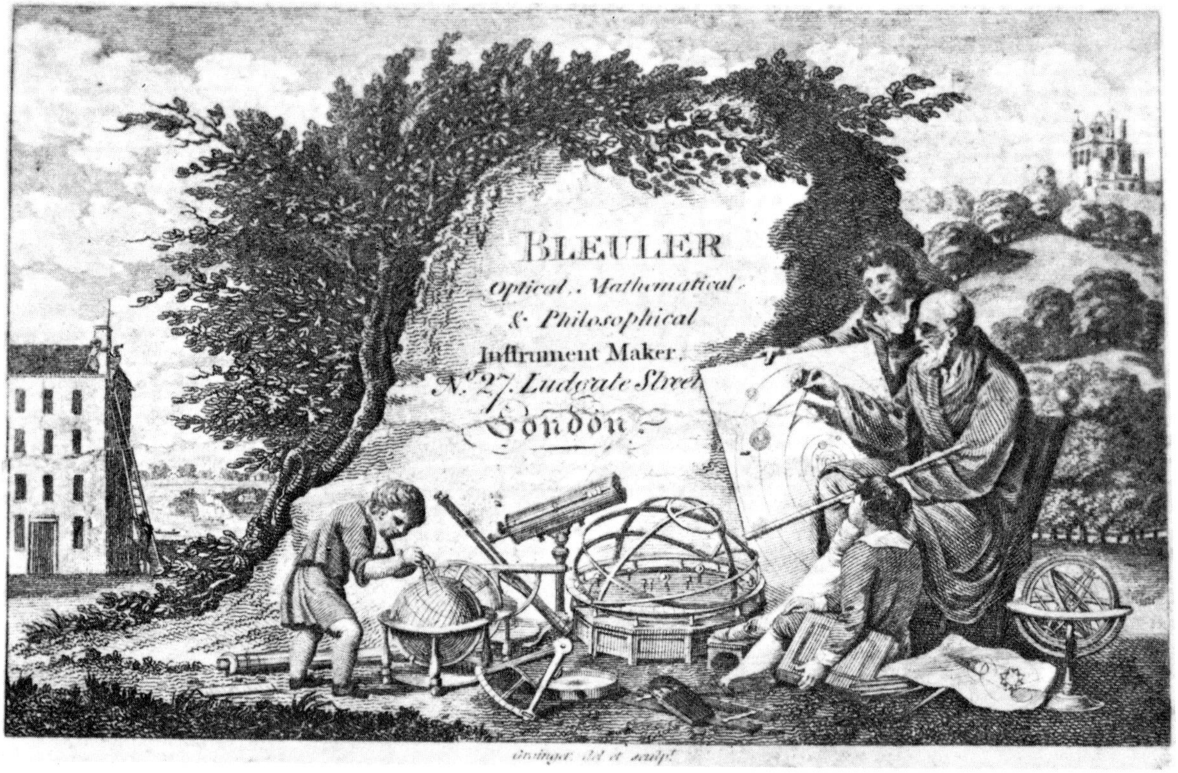

BLEULER
Optical, Mathematical,
& Philosophical
Instrument Maker,
No. 27. Ludgate Street
London

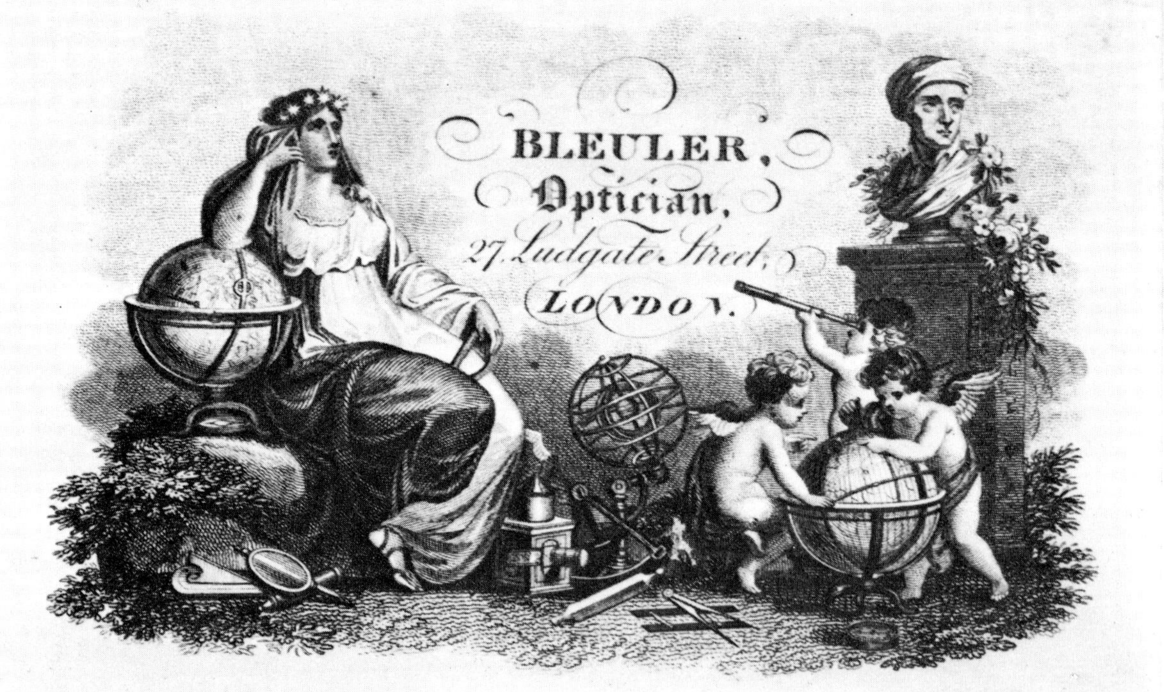

BLEULER,
Optician,
27. Ludgate Street,
LONDON.

top left

11 John Bleuler
catalogue no.52

bottom left

12 John Bleuler
catalogue no.53

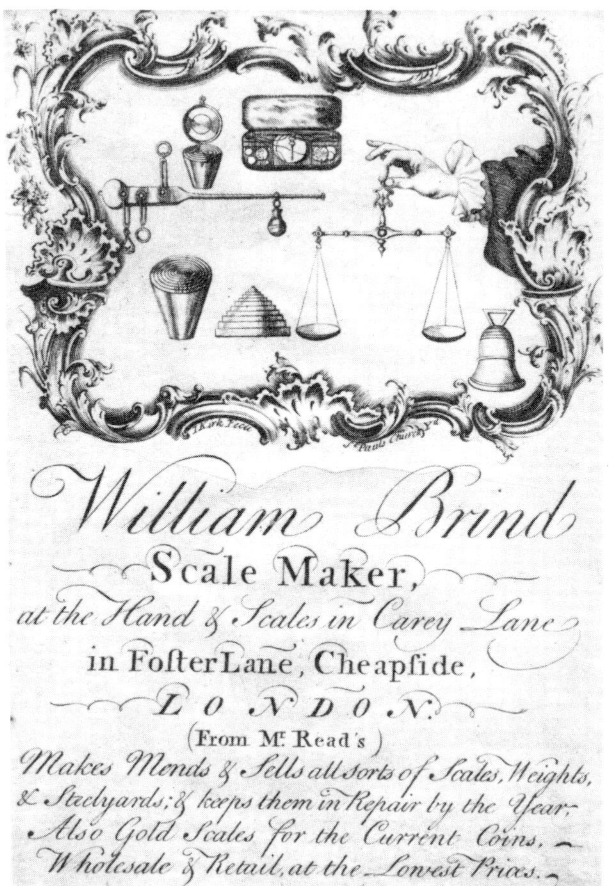

13 William Brind
catalogue no.61

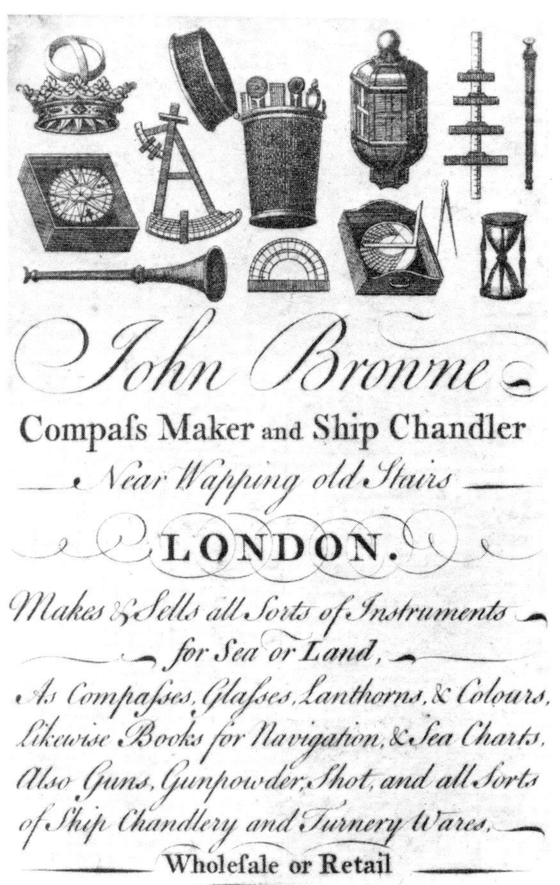

14 John Browne
catalogue no.64

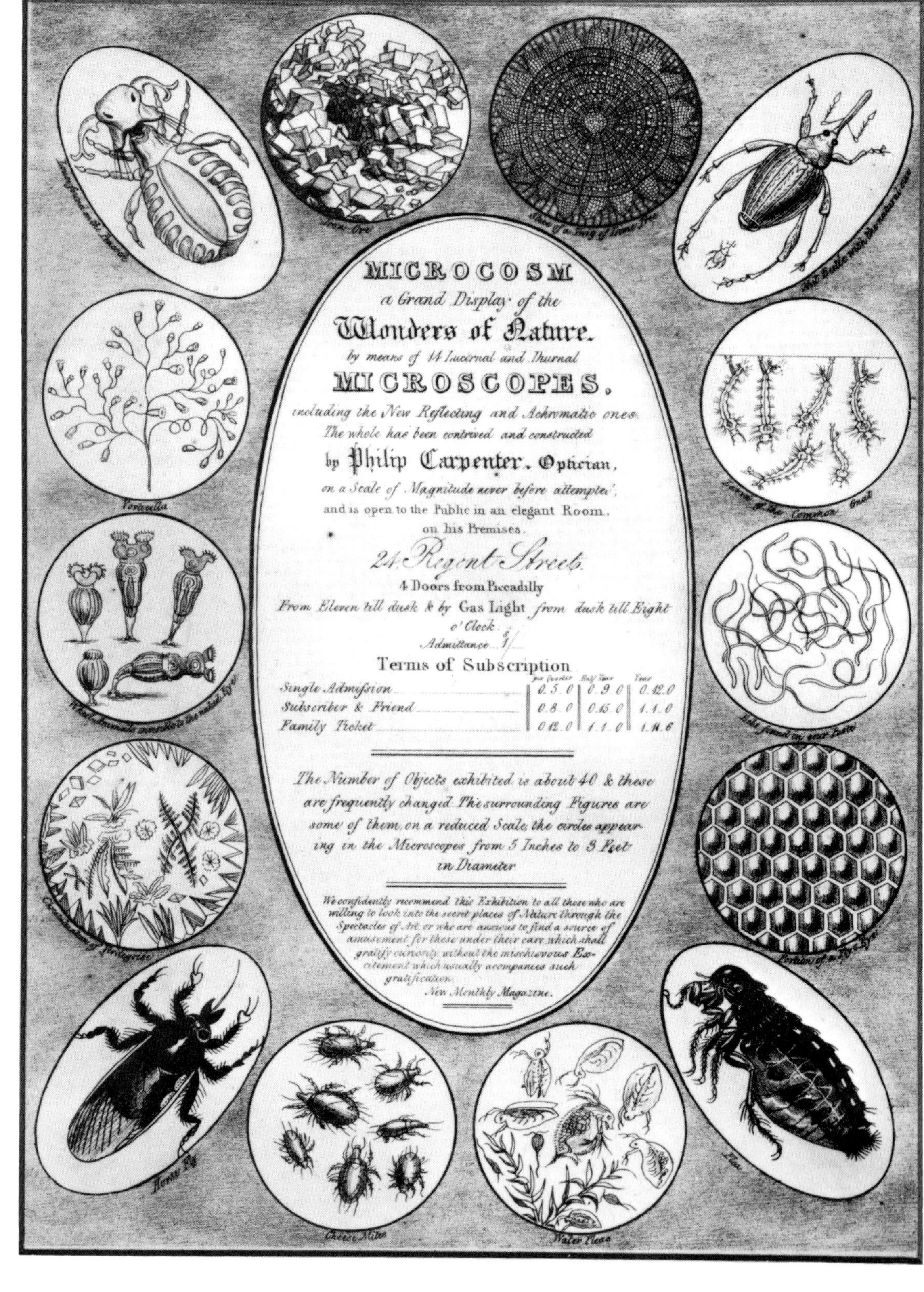

MICROCOSM
a Grand Display of the
Wonders of Nature.
by means of 14 Lucernal and Diurnal
MICROSCOPES.
including the New Reflecting and Achromatic ones.
The whole has been contrived and constructed
by Philip Carpenter. Optician,
on a Scale of Magnitude never before attempted,
and is open to the Public in an elegant Room,
on his Premises.
24, Regent Street.
4 Doors from Piccadilly
From Eleven till dusk & by Gas Light from dusk till Eight
o'Clock.
Admittance 1/
Terms of Subscription
Single Admission — 0.5.0 — 0.9.0 — 0.12.0
Subscriber & Friend — 0.8.0 — 0.15.0 — 1.1.0
Family Ticket — 0.12.0 — 1.1.0 — 1.11.6
The Number of Objects exhibited is about 40 & these
are frequently changed. The surrounding Figures are
some of them, on a reduced Scale, the circles appear-
ing in the Microscopes from 5 Inches to 3 Feet
in Diameter
We confidently recommend this Exhibition to all those who are
willing to look into the secret places of Nature through the
Spectacles of Art, or who are anxious to find a source of
amusement for those under their care, which shall
gratify curiosity, without the mischievous Ex-
citement which usually accompanies such
gratification.
New Monthly Magazine.
Vorticella
Wheel Animalcula invisible to the naked Eye
Crystallization of Phlogistic
Honey Fly
Cheese Mites
Water Fleas
Flea
Portion of a Flys Eye
Eels found in sour Paste
Larvae of the Common Gnat
The Nut, or the Nut-Weevil
Slice of a Twig of Oak
Iron Ore
Gizzard with the Teeth

left
15 Philip Carpenter
catalogue no.71

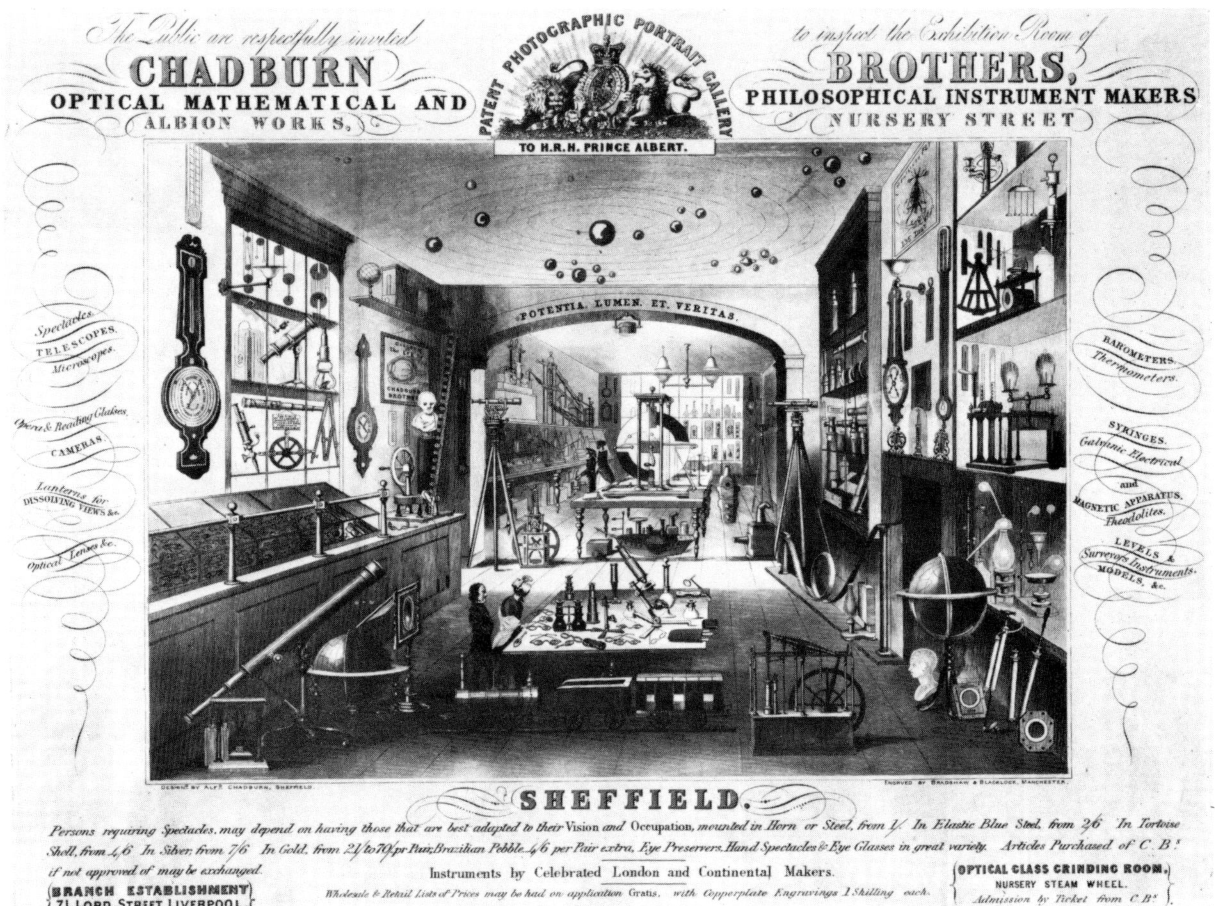
The Public are respectfully invited to inspect the Exhibition Room of
CHADBURN BROTHERS,
OPTICAL MATHEMATICAL AND PHILOSOPHICAL INSTRUMENT MAKERS
ALBION WORKS, NURSERY STREET
PATENT PHOTOGRAPHIC PORTRAIT GALLERY
TO H.R.H. PRINCE ALBERT.
POTENTIA, LUMEN, ET VERITAS.
Spectacles.
TELESCOPES.
Microscopes.
Opera & Reading Glasses.
CAMERAS
Lanterns for DISSOLVING VIEWS &c.
Optical Lenses &c.
BAROMETERS.
Thermometers.
SYRINGES.
Galvanic Electrical
and
MAGNETIC APPARATUS.
Theodolites.
LEVELS &
Surveyors Instruments.
MODELS. &c.
SHEFFIELD.
Persons requiring Spectacles, may depend on having those that are best adapted to their Vision and Occupation, mounted in Horn or Steel, from 1/. In Elastic Blue Steel, from 2/6. In Tortoise Shell, from 4/6. In Silver: from 7/6. In Gold, from 21/ to 70/ pr Pair, Brazilian Pebble 4/6 per Pair extra, Eye Preservers, Hand Spectacles & Eye Glasses in great variety.   Articles Purchased of C.B.S if not approved of may be exchanged.
Instruments by Celebrated London and Continental Makers.
OPTICAL GLASS GRINDING ROOM.
NURSERY STEAM WHEEL.
BRANCH ESTABLISHMENT
71 LORD STREET, LIVERPOOL.
Wholesale & Retail Lists of Prices may be had on application Gratis, with Copperplate Engravings 1 Shilling each.
Admission by Ticket from C.B.S

above
16 Chadburn Brothers
catalogue no.83

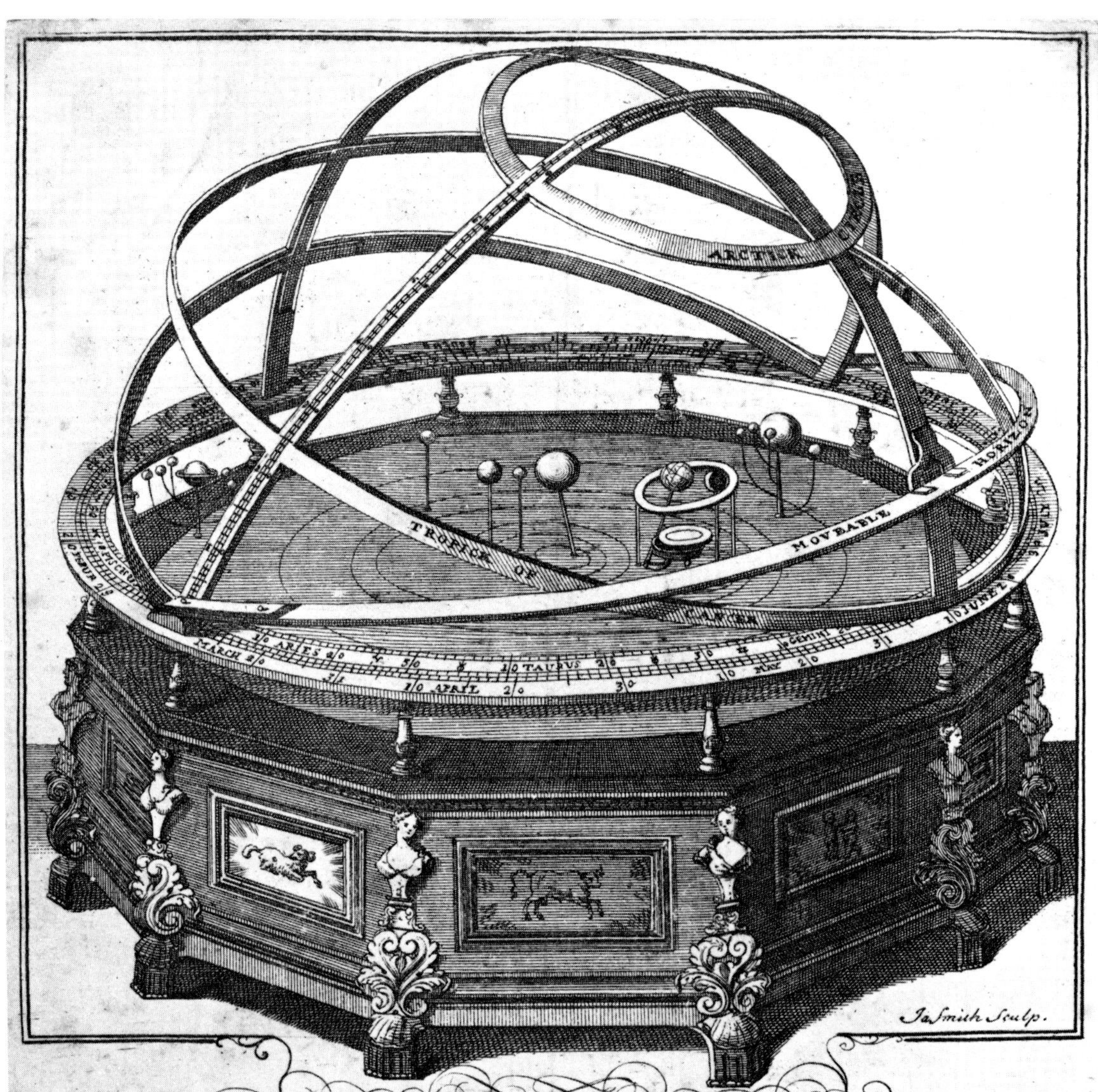

The Grand Orrery

And all other Mathematical
Instruments made and sold by

Benja:N Cole,

at y^e Royal Exchange, or at his House in Ball Alley,
going out of George-Yard, into Lombard Street.

left

17 Benjamin Cole
catalogue no.91

right

18 Colombi
catalogue no.94

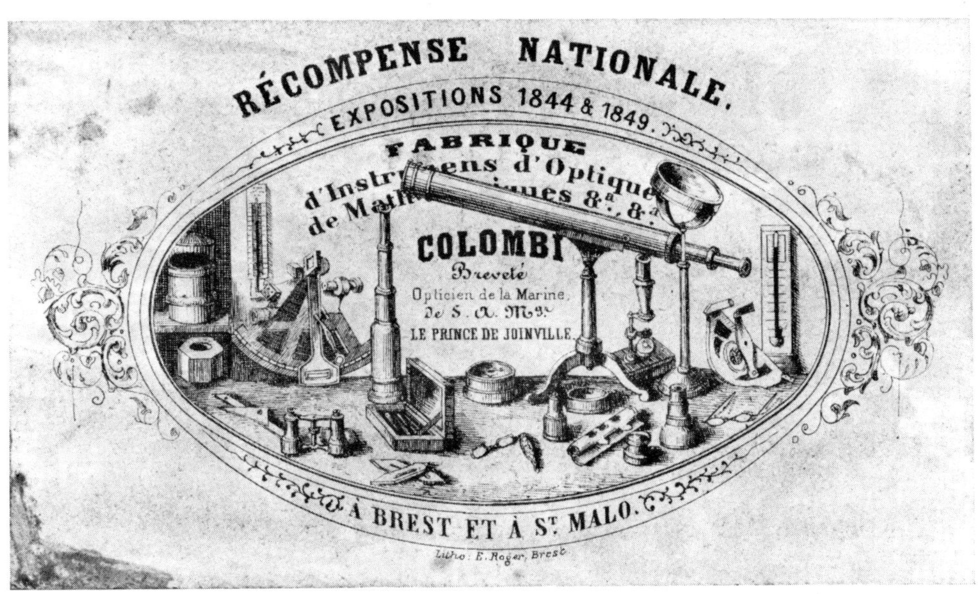

right

19 Fisher Combes
catalogue no.95

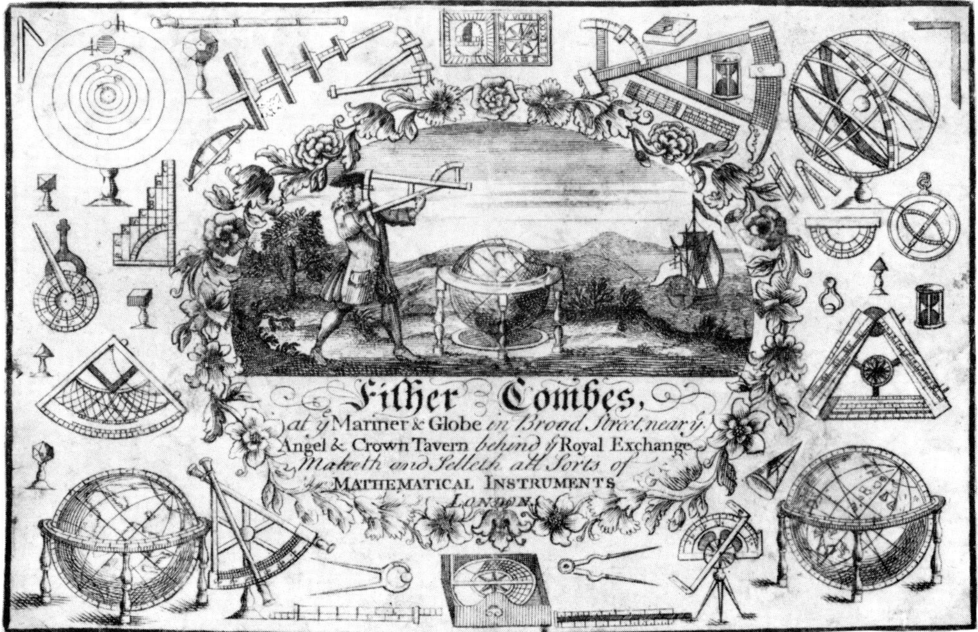

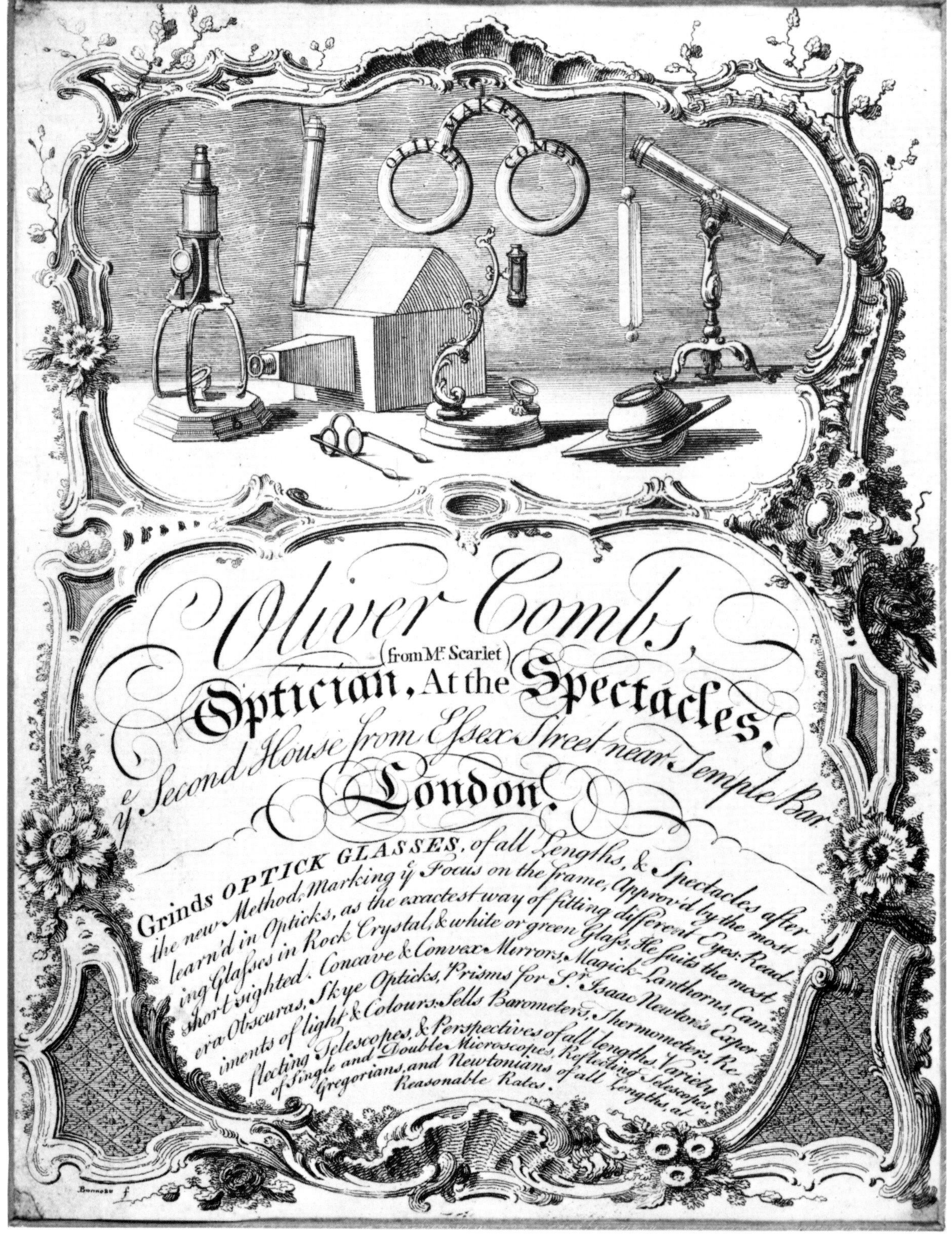

OPTICUM MAKER COMBS

Oliver Combs,
(from Mr. Scarlet)
Optician, At the Spectacles,
ye Second House from Essex Street near Temple Bar,
London.

Grinds OPTICK GLASSES, of all Lengths, & Spectacles after
the new Method, Marking ye Focus on the frame, Approv'd by the most
learn'd in Opticks, as the exactest way of fitting different Eyes: Reading Glasses in Rock Crystal, & white or green Glass, He suits the most
short-sighted. Concave & Convax Mirrors, Magick Lanthorns, Camera Obscuras, Skye Opticks, Prisms for Sr. Isaac Newton's Experiments of light & Colours, Sells Barometers, Thermometers, Reflecting Telescopes, & Perspectives of all lengths, Variety
of Single and Double Microscopes, Reflecting Telescopes,
Gregorians, and Newtonians of all Lengths, at
Reasonable Rates.

Bennett f.

left

20 Oliver Combs
catalogue no. 96

right

21 John Cuff
catalogue no.101

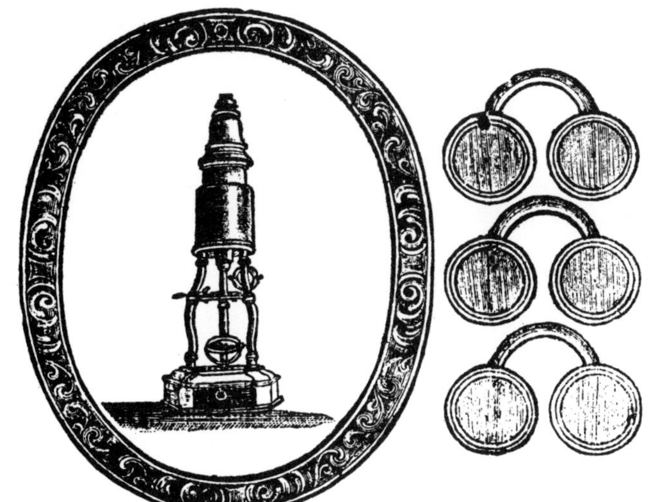

J O H N   C U F F,

Optician, Spectacle, *and* Microscope *Maker,*

At the Sign of the REFLECTING MICROSCOPE and SPECTACLES,
againſt *Serjeant's-Inn* Gate in *Fleet-ſtreet,*

MAKES and Sells all Sorts of the moſt curious Optical Inſtruments, ſuch as SPEC-TACLES and READING GLASSES, ground on Braſs Tools in the Method approved by the Royal Society, of *Brazil* Pebbles, Rock Chryſtal, or the fineſt Flint Glaſs, and ſet either in Silver, Mother of Pearl, Tortoiſe-ſhell, *&c,* in the moſt convenient Manner for Uſe.

REFRACTING TELESCOPES, proper for celeſtial or terreſtrial Obſervations, either by Land or Sea; and alſo REFLECTING TELESCOPES, as invented by Sir *Iſaac Newton* and Mr *Gregory.*

MICROSCOPES of ſeveral Kinds, both Single and Double, (either for the Study or the Pocket) particularly a new-conſtructed DOUBLE MICROSCOPE, invented by the ſaid *John Cuff,* to remedy the Inconveniencies complained of in all the former Sorts; the SOLAR MICROSCOPE, as improv'd by him with a moſt convenient and manageable Apparatus for viewing Objects, and a Contrivance for drawing the Pictures of them with great Eaſe and Exactneſs, even by thoſe that have never drawn before; the AQUATIC MICROSCOPE, invented alſo by him for the Examination of Water Animals; the OPAKE MICROSCOPE, which by a Light reflected upon Objects that have no Tranſparency, renders them diſtinctly viſible; *Culpepper's* MICROSCOPE, *Wilſon's* Pocket MICROSCOPE, *&c.*

The CAMERA OBSCURA for exhibiting Proſpects in their natural Proportions and Colours, together with the Motions of living Subjects: MAGIC LANTHORNS; Convex and Concave SPECULUMS or LOOKING GLASSES, MULTIPLYING GLASSES, BAROMETERS, THERMOMETERS, OPERA GLASSES, SPECTACLES of true *Venetian* Green Glaſs, SPEAKING TRUMPETS, ſeveral Mathematical Inſtruments, and many other Curioſities, are ſold by the ſaid Operator at reaſonable Prices, either Wholeſale or Retale.

below
22 E. Culpeper
    catalogue no.106

right
23 W. Dowling
    catalogue no.132

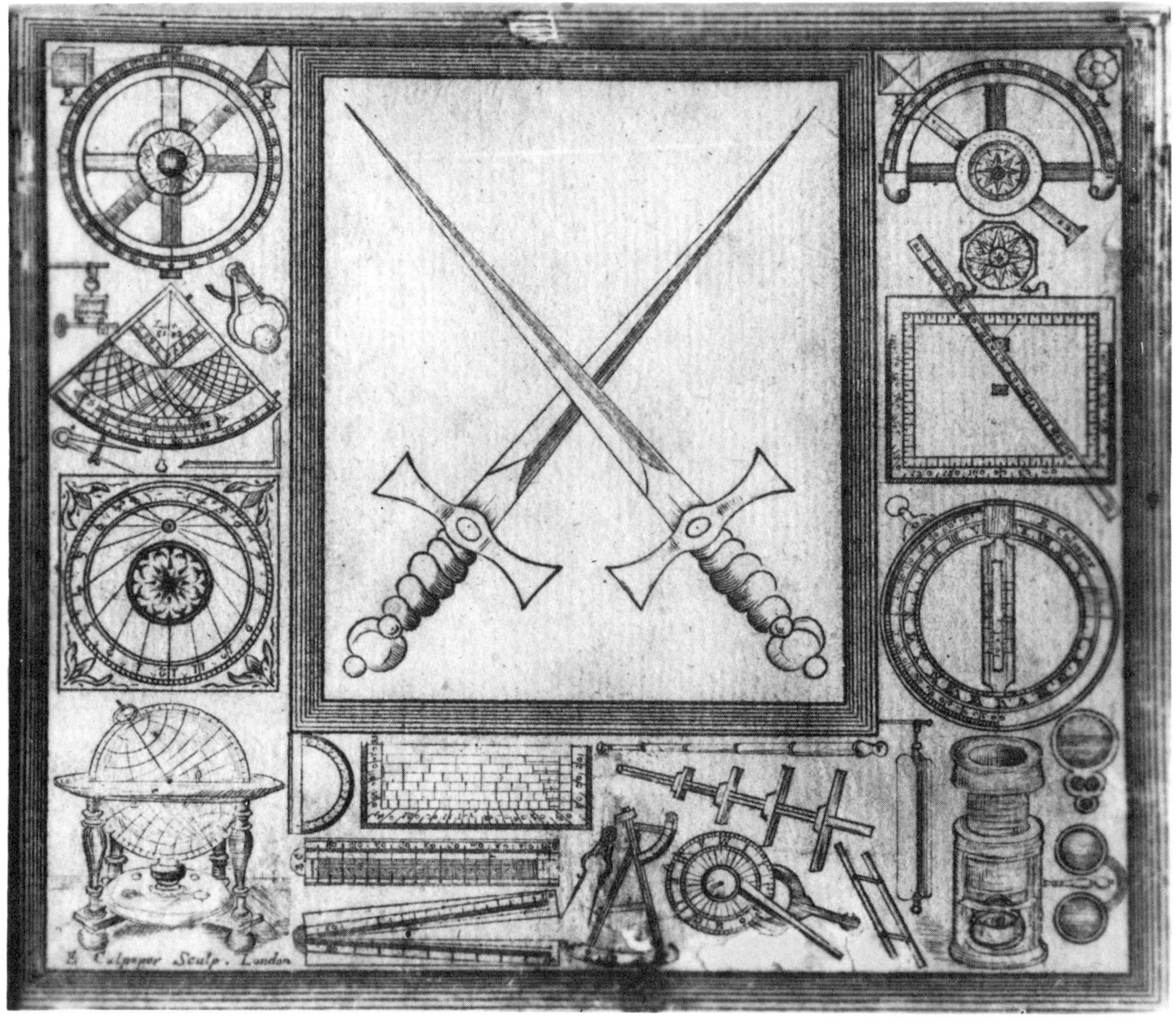

E. Culpeper Sculp. London

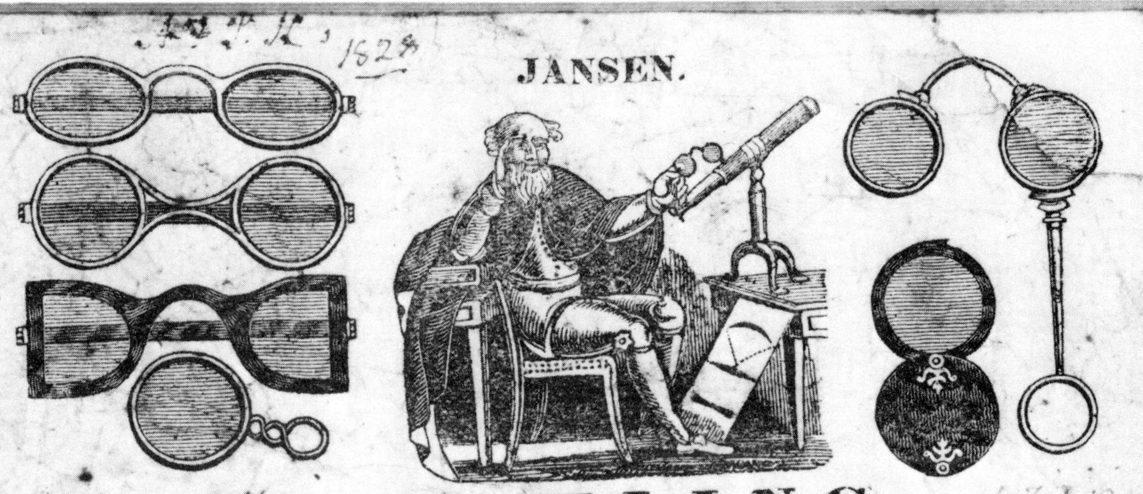

W. DOWLING,

WORKING OPTICIAN,

WEST GATEWAY of LINCOLN'S-INN, SERLE STREET, LINCOLN'S-INN-FIELDS, LONDON;

Manufactures and Sells all Kinds of Optical, Mathematical, & Philosophical Instruments, of the best Workmanship, and according to the latest Improvements.

SPECTACLES for long and short-sighted Eyes, mounted in Gold, Silver, Tortoiseshell, and Steel, with the finest White or coloured Glasses, or with the best Brazil Pebbles, which will not scratch nor injure by wearing. Spectacles for short-sighted Persons in a newly contrived mounting. These Frames cannot be displaced from the Head by the motion of the wearer, and are therefore peculiarly calculated for the Officers of the Army and Navy, and Sportsmen. Improved Spectacles, for Persons whose Eyes have undergone the operation of Couching, or other surgical process, or whose Eyes are much enfeebled from the effects of sickness, or warm climates, &c.

In these new Spectacles there are some satisfactory improvements in the construction of the Frames; but the colour of the Glasses, and the mode of distributing that colour over the field of view, is completely new, and is founded on simple optical facts, which have hitherto escaped the notice of Spectacle Makers.

Tortoiseshell Spectacles, peculiar for their lightness and uninterruption of dressed hair. Spectacles with Shade, to secure weak and inflamed eyes from too intense a light. New Reading Spectacles for the Theatres, Public Lectures, &c. which fold up and may be suspended by a ribbon. Spectacles of every description, to counteract the effects of age, mounted in Gold, Silver, Tortoiseshell or Steel. New Walking or Riding Spectacles, for the protection of weak eyes. New Spectacles, to protect the eyes against the Sun, Wind, Dust, Snow, Sleet, &c. or an excess of Light under any circumstances. New coloured Spectacles for Artists, to facilitate sketching and tinting from Nature.

Concave Glasses in great variety, for short sighted Eyes. Preservers for Eyes first requiring the aid of Spectacle. Reading and Magnifying Glasses, for all the variety of uses, mounted in Gold, Silver, Pearl, and Tortoiseshell, &c.

The Glasses or Brazil Pebbles manufactured by W. DOWLING are ground in accurate tools of concave or convex or spherical figures, under HIS OWN HANDS; having been apprenticed to that particular branch, he cannot fail of suiting every description of weak or impaired sight. One particular property in which his Spectacles excel, is that of having the Glasses or Brazil Pebbles centered by an engine made for that purpose; by which means, the centre of the Glass is brought in contact with the pupil of the eye, and thereby exhibits objects clear and distinct.

Those Persons, who may please to favor W. D. with a trial, may depend on being carefully suited with the best Glasses or Brazil Pebbles, properly adapted to their sight, for the Eyes of numbers of Persons have been greatly injured by bad or improper choice of Spectacles, in buying them from Hawkers, Pedlars, and others, who are unacquainted with the nature of Vision, deal in a sort of Spectacles too bad for any but themselves to recommend, and are unable to adopt properly various Glasses to different eyes. The Glasses may be exchanged after a short trial if required, provided they do not suit. Persons in the country or abroad, may be exactly suited by sending the Glasses or part of the ones they last used; and those not having used Glasses, may be suited by sending word whether they have ever worn Spectacles, and at what distance, in inches, they can or cannot best discern the smallest print of this paper, without the aid of Spectacles.

Improved Opera Glasses, both Plain and Achromatic, so very desirable in a Theatre, in various mountings.—*Refracting Telescopes*, for Land or Sea.—*Reflecting Telescopes*.—*Perspective Glasses*.—*Single and Compound Microscopes*.—*Camera Obscurae*.—*Diagonal Optical Machines*, for viewing Prints.—*Concave and Convex Mirrors*.—*Prisms*.—*Magic Lanterns* with a variety of Slides.—*Barometers* and *Thermometers*.—*Quadrants and Sextants*.—*Globes*, in various sizes.—*Electrical Machines*.—*Air Pumps*.—*Pentegraphs*.—*Theodolites*.—*Circumferenters*.—*Levels*.—*Measuring Chains* and *Tapes*.—*Drawing Instruments*.—*Gauging and Plotting Rules* and *Scales*.—*Protractors* and *Parallel Rules*, together with every Instrument in general request.

N. B.—*Spectacles Repaired*, and Glasses suited, or Pebbles re-ground to suit the Sight of the wearer. *Jewellers supplied with Glasses to suit all Sights.*

REAL GOLCONDA LOADSTONE, A SURE PREVENTIVE FOR GOUT AND ACUTE RHEUMATISM. Second-hand Instruments Bought, Sold, or Exchanged.

R. Wilks, Printer, 89, Chancery Lane.

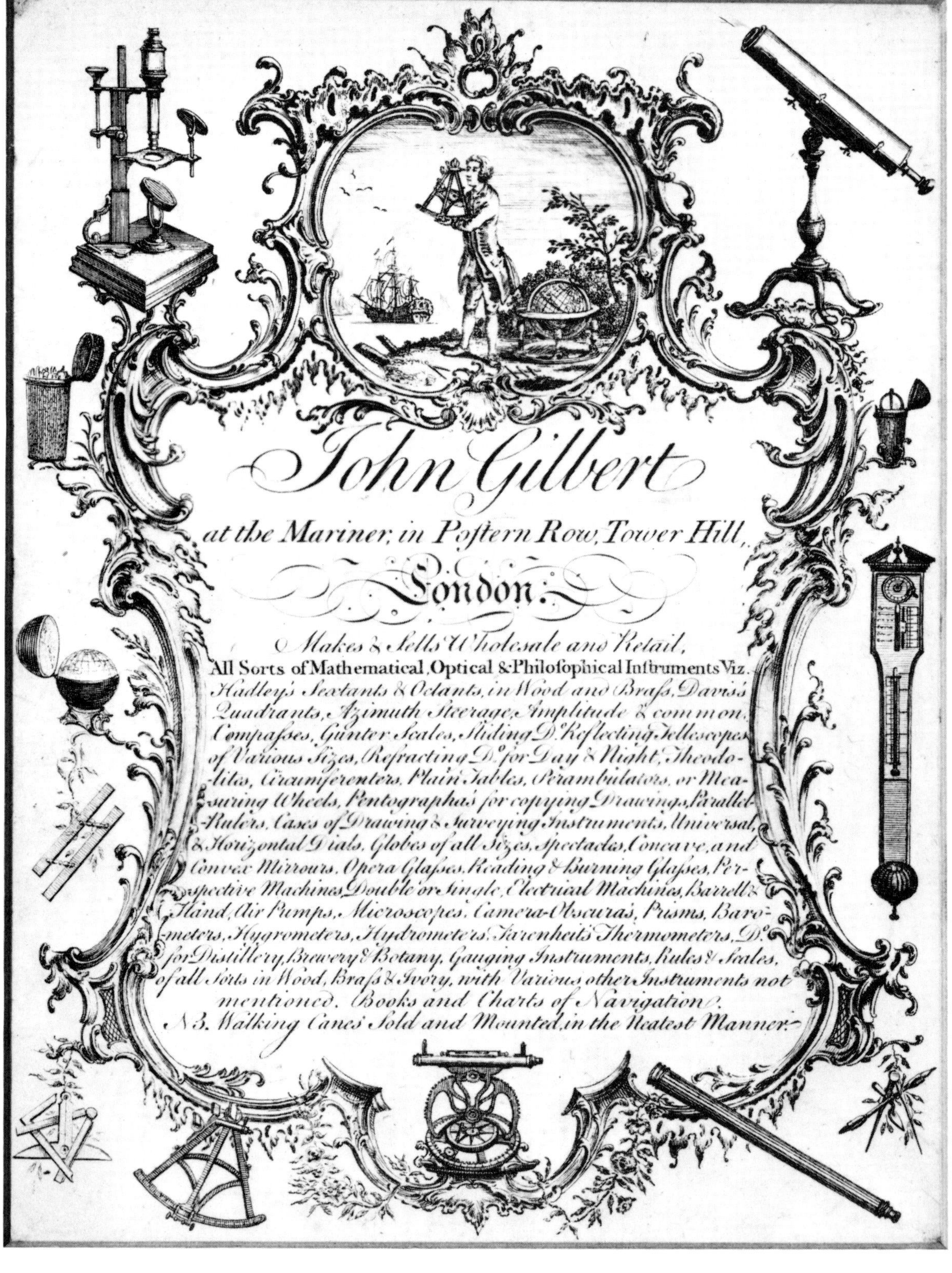
John Gilbert
at the Mariner, in Postern Row, Tower Hill,
London.
Makes & Sells Wholesale and Retail,
All Sorts of Mathematical, Optical & Philosophical Instruments Viz.
Hadley's Sextants & Octants, in Wood and Brass, Davis's
Quadrants, Azimuth Steerage, Amplitude & common
Compasses, Gunter Scales, Sliding Dº Reflecting Tellescopes
of Various Sizes, Refracting Dº for Day & Night, Theodo-
lites, Circumferenters, Plain Tables, Perambulators, or Mea-
suring Wheels, Pentographes for copying Drawings, Parallel
Rulers, Cases of Drawing & Surveying Instruments, Universal
& Horizontal Dials, Globes of all Sizes, Spectacles, Concave, and
Convex Mirrours, Opera Glasses, Reading & Burning Glasses, Perspective Machines Double or Single, Electrical Machines, Barrell &
Hand, Air Pumps, Microscopes, Camera-Obscuras, Prisms, Barometers, Hygrometers, Hydrometers, Farenheits Thermometers, Dº
for Distillery, Brewery & Botany, Gauging Instruments, Rules & Scales,
of all Sorts in Wood, Brass & Ivory, with Various other Instruments not
mentioned. Books and Charts of Navigation.
N.B. Walking Canes Sold and Mounted, in the Neatest Manner.

left
24 John Gilbert
catalogue No.156

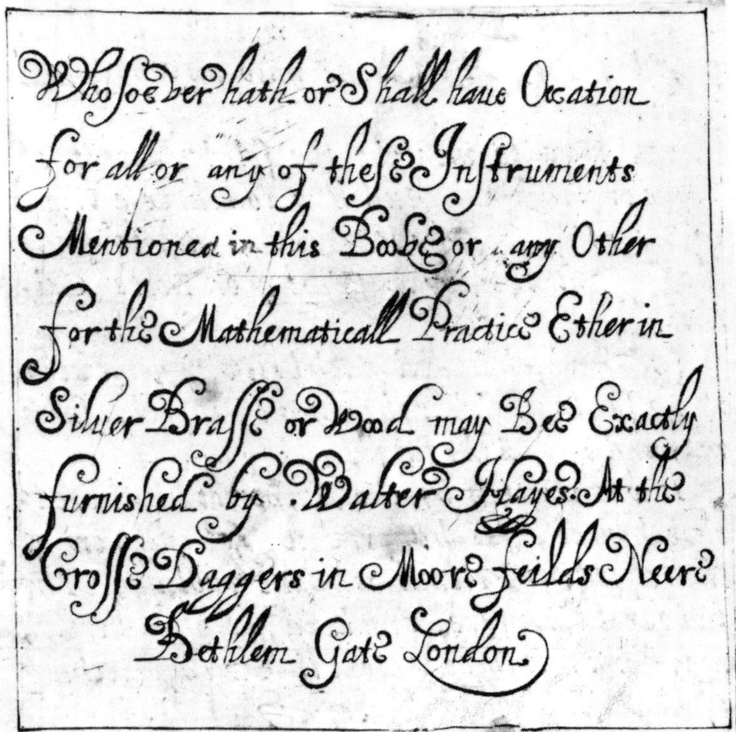
Whosoever hath or Shall have Occation
for all or any of these Instruments
Mentioned in this Booke or any Other
for the Mathematicall Practice Either in
Silver Brass or Wood may Bee Exactly
furnished by Walter Hayes At the
Crosse Daggers in Moore feilds Neere
Bethlem Gate London

right
25 Walter Hayes
catalogue no.182

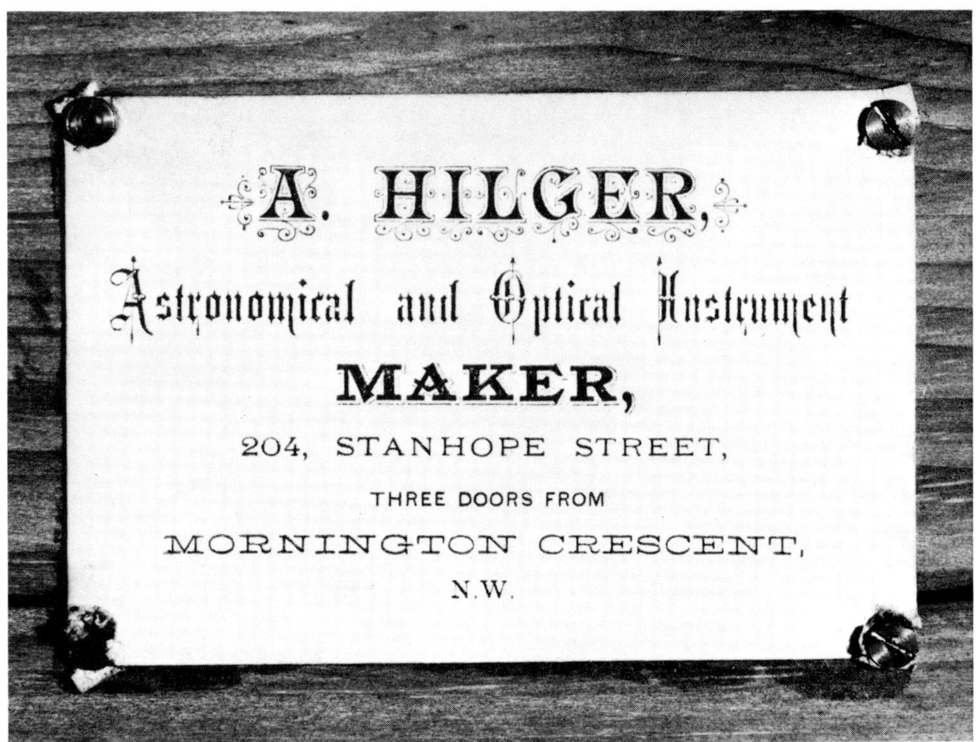
A. HILGER,
Astronomical and Optical Instrument
MAKER,
204, STANHOPE STREET,
THREE DOORS FROM
MORNINGTON CRESCENT,
N.W.

right
26 A. Hilger
catalogue no.193

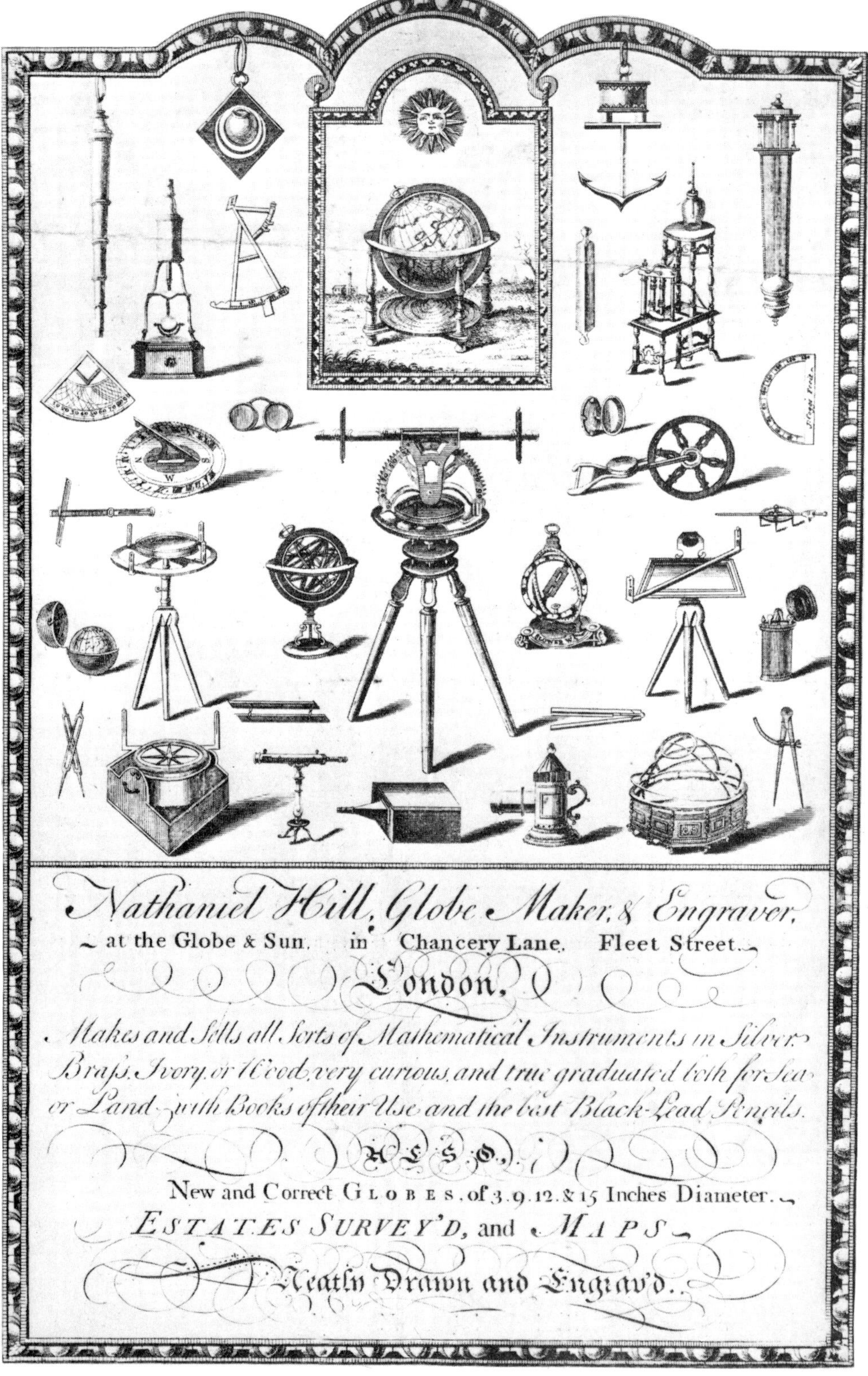

left
27 Nathaniel Hill
catalogue no. 194

right
28 Samuel Johnson
catalogue no.209

right
29 W. and S. Jones
catalogue no.219

[THE OLDEST SHOP.]

SAMUEL JOHNSON,

OPTICIAN,

(Successor to the late Mr. *MANN*)

At the Sign of Sir Isaac Newton and Two Pair of Golden Spectacles, near the West End of St. *Paul's*, London;

MAKES and SELLS, Wholesale and Retail, the finest Crystal Spectacles, ground upon Brass Tools (approved by the ROYAL SOCIETY and the greatest Mathematicians) set in Gold, Silver, Tortoise-shell, Horn, Leather, and all Manner of Ways that are most convenient and beautiful. Also curious Reading Glasses of Rock Crystal, or the whitest Flint, contrived to turn into Cases of various Kinds, to be carried in the Pocket without Trouble or Damage. Likewise Spectacles of *Brazil*-Pebble, or Green and Blue Glass.

He makes Concaves for Myops or Short-sighted Persons, which are so nicely adapted to every Degree of Short-sightedness, and by so regular a Method, that Persons, after being once fitted, may at all Times be furnished with them (though at ever so great a Distance) as exact as if present.

REFLECTING TELESCOPES on the Principles of Sir Isaac Newton, and Mr. Gregory, made with the greatest Accuracy, and the Apparatus finished and adapted in so compleat a Manner, as renders the Use of them pleasant and easy. Also MICROSCOPES double and single (with all the modern Improvements) which magnify to so great a Degree, that the Circulation of the Blood in Animals, the Animalcula in Fluids, and the Farina of Vegetables, with many other Phænomena, otherwise imperceptible, are clearly discover'd.
———— Particularly an Improvement of the Apparatus of the Solar Microscope, which renders it much easier for Use, and less liable to Disorder, than any yet extant.

PRISMS for demonstrating the surprizing Theory of Light and Colours; Camera Obscura's for delineating Views in Perspective; Optical Machines for viewing Perspective Pictures; Convex and Concave Speculums of all Sizes; Magick Lanthorns, Opera Glasses, Multiplying and Magnifying Glasses, Barometers and Thermometers; with many other Curiosities not here mentioned.

REFRACTING TELESCOPES IMPROVED,

By a Method founded on the justest Principles and Rules of Opticks, are allowed by the best Judges in Theory (who have made the Comparison) to be equal in all Respects to those of the celebrated PETRO PATRONE at *Milan*; by which Improvement, those of two, three, and four Feet, have by Experience been found to be much more useful at Sea, than any have hitherto been, and are made by the abovesaid SAMUEL JOHNSON.

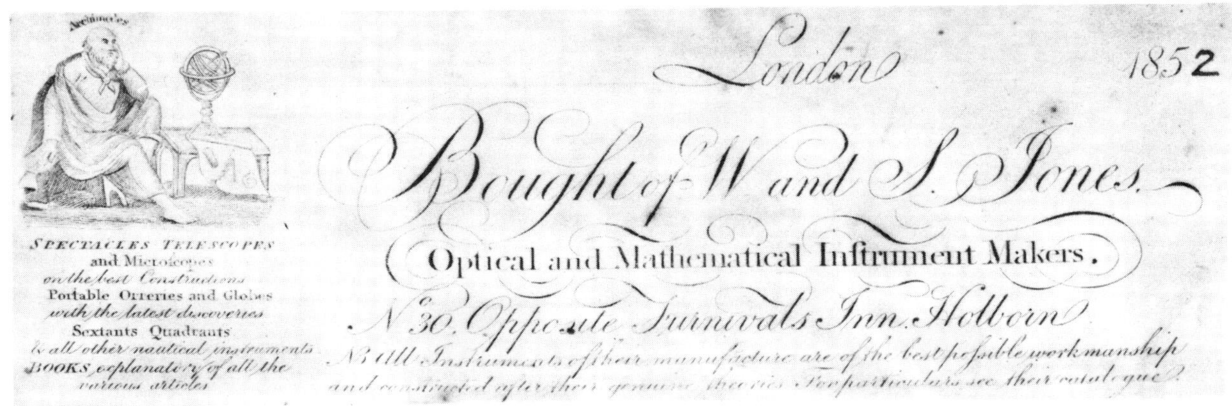

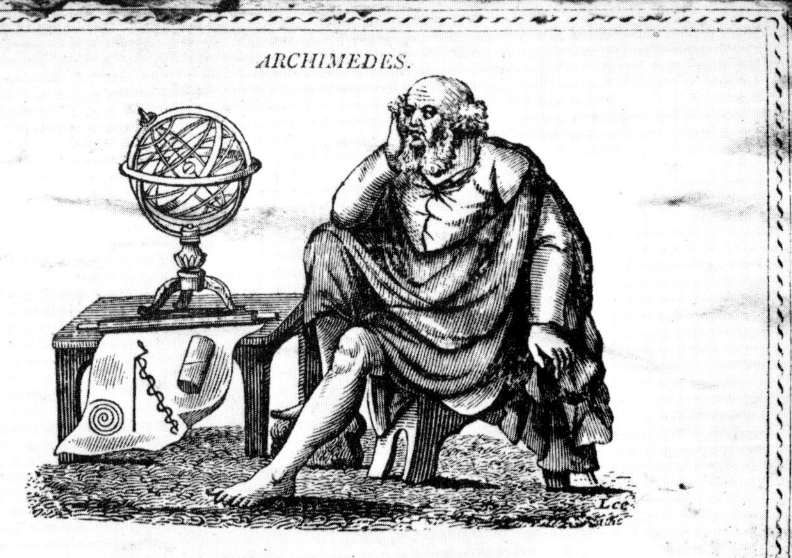

left
30 W. and S. Jones
catalogue no.221

right
31 George Lee and Son
catalogue no.233

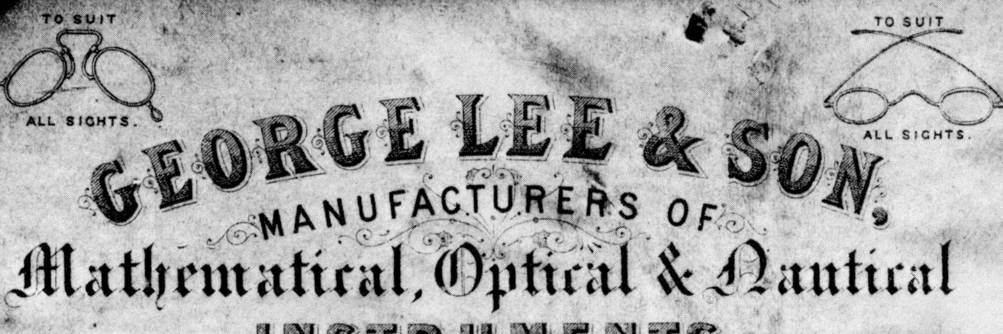

TO SUIT
ALL SIGHTS.
TO SUIT
ALL SIGHTS.
GEORGE LEE & SON,
MANUFACTURERS OF
Mathematical, Optical & Nautical
INSTRUMENTS,
TO THE HON.ble CORPORATION OF THE
TRINITY HOUSE & THE ADMIRALTY.
ORDNANCE ROW,
THE HARD, PORTSEA,
AND 3, PALMERSTON Rd, SOUTHSEA.

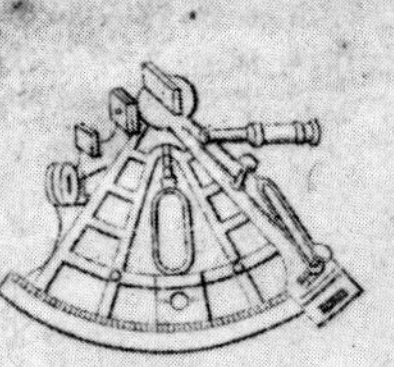
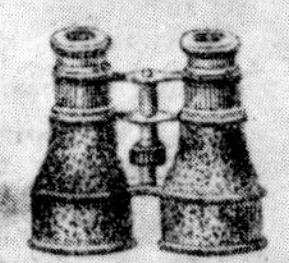

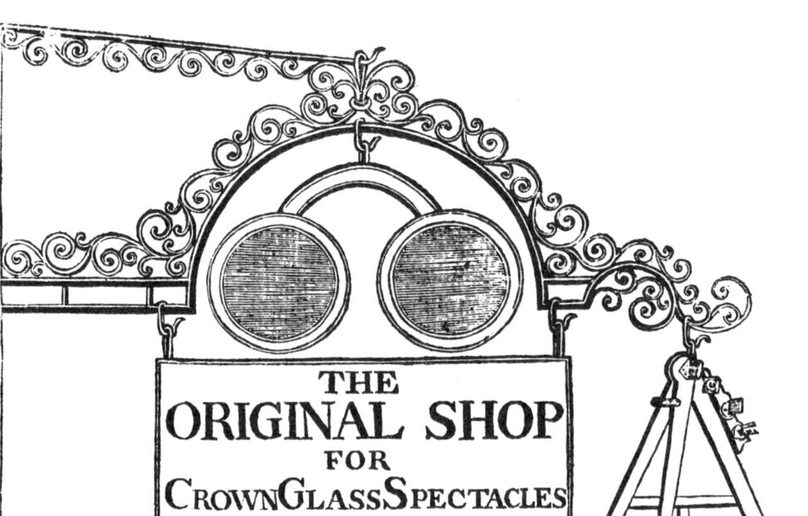

JOSEPH LINNELL,

(Succeſsor to the late Mr. *JAMES AYSCOUGH,*)

OPTICIAN,

At the Great GOLDEN SPECTACLES and QUADRANT, Nº. 33, in *Ludgate-Street,* near St. PAUL's, *LONDON,*

(*The Original Shop for ſuperfine* CROWN-GLASS SPECTACLES.)

MAKES and Sells (Wholeſale and Retail) SPECTACLES and READING-GLASSES, either of *Brazil* Pebbles, Crown, White, Green, or Blue Glaſs, ground after the trueſt Method, ſet in neat and commodious Frames, ſome of which are ſo contrived as to preſs neither upon the Noſe nor the Temples.

CONCAVES for Short-ſighted Perſons.

MICROSCOPES of various Kinds, particularly one of a new Conſtruction, which is uſed either as a ſingle one or a double one, for tranſparent or opake Objects; the Magnifiers of which are adapted for viewing the ſmalleſt Object in Nature, as alſo thoſe of the Size of a common Flower. It may be uſed with the Solar Apparatus.

REFLECTING TELESCOPES of all Sizes.

REFRACTING TELESCOPES of various Lengths, ſome of which are peculiarly adapted to uſe at Sea, and are ſo contrived to be adjuſted in ſuch a Manner as to fit any Eye, whether ſhort-ſighted, or the reverſe. To which alſo is added a Conveniency for receiving or excluding more or leſs Light, according as the Brightneſs or Cloudineſs of the Atmoſphere requires. Which by ſeveral late Trials at Sea were found very advantageous. Alſo, a New-invented Achromatic TELESCOPE, which magnify much more, than thoſe made on the common Principle.

PRISIS for demonſtrating the Theory of Light and Colours. Scioptic BALLS.

CAMERA OBSCURA's for delineating Landſkips and Proſpects, Concave and Convex SPECULUMS, MAGICK-LANTHORNS, OPERA-GLASSES, &c.

OPTICAL MACHINES for viewing Perſpective Prints.

BAROMETERS, Diagonal, Standard, or Portable.

THERMOMETERS, whoſe Scales are adjuſted to the Bores of their reſpective Tubes; Hygrometers, and Hydrometers; Hydroſtatical Balances; *Hadley's* Quadrants; Globes; Caſes of Drawing Inſtruments; Scales; Parallel Rulers; Sun-Dials; Rules; Pencils; with all other Sorts of Optical, Mathematical, or Philoſophical Inſtruments.

left

32 Joseph Linnell
catalogue no.239

right

33 James Mann
catalogue no.251

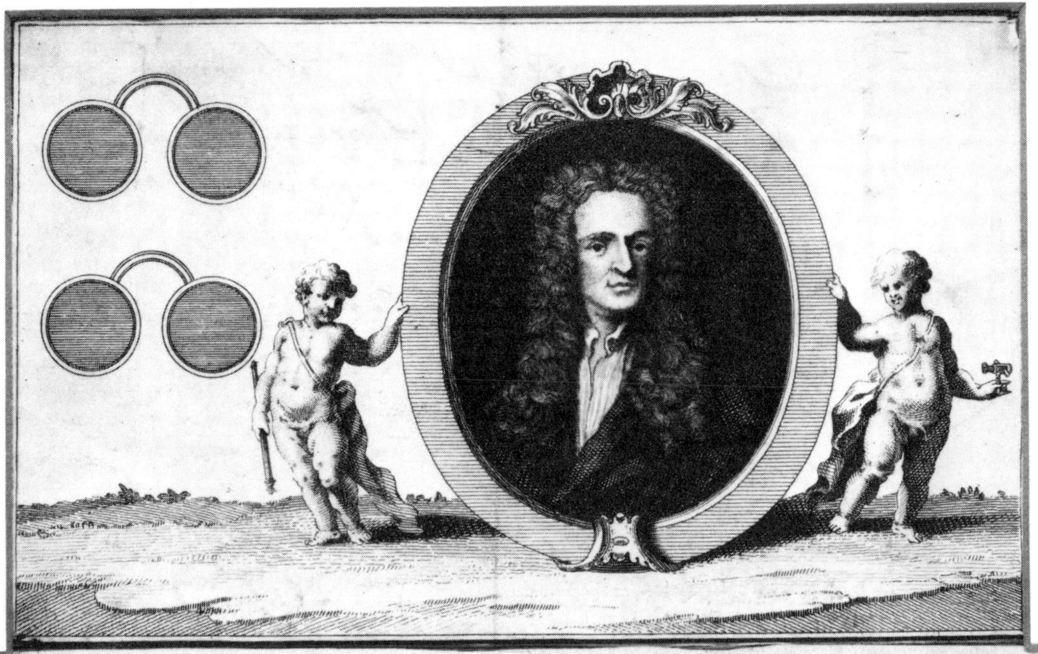

JAMES MANN,

OPTICIAN,

At the Sign of Sir *Isaac Newton*, and Two Pair of Golden Spectacles, near the West End of St. *Paul*'s, LONDON.

MAKES and SELLS, Wholesale and Retail, the finest Crystal Spectacles, ground upon Brass Tools, (approved by the *Royal Society* and the greatest Mathematicians) set in Gold, Silver, Tortoiseshell, Horn, Leather, and all Manner of Ways that are most convenient and beautiful. Also curious Reading-Glasses of Rock Crystal, or the whitest Flint, contrived to turn into Cases of various Kinds, to be carried in the Pocket without Trouble or Damage. Likewise Spectacles of *Brazil*-Pebble, or Green and Blue Glass.

N. B. Whereas false-ground Spectacles are very prejudicial to the Eyes; the Curious may here be shewn, in a Minute's Time, how to distinguish between those that are true and those that are false.

He makes Concaves for Myops or Short-sighted Persons, which are so nicely adapted to every Degree of Short-sightedness, and by so regular a Method, that Persons, after being once fitted, may at all Times be furnished with them (though at ever so great a Distance) as exact as if present.

Reflecting Telescopes on the Principles of Sir *Isaac Newton* and Mr. *Gregory*, made with the greatest Accuracy, and the Apparatus finished and adapted in so compleat a Manner, as renders the Use of them pleasant and easy. Also Microscopes double and single, with all the modern Improvements, which magnify to so great a Degree, that the Circulation of the Blood in Animals, the Animalcula in Fluids, and the Farina of Vegetables, with many other Phænomena, otherwise imperceptible, are clearly discover'd. ——Particularly an Improvement of the Apparatus of the Solar Microscope, which renders it much easier for Use, and less liable to Disorder, than any yet extant.

Prisms for demonstrating the surprizing Theory of Light and Colours; Camera Obscuras for delineating Views in Perspective; Optical Machines for viewing Perspective Pictures; Convex and Concave Speculums of all Sizes; Magick Lanthorns, Opera Glasses, Multiplying and Magnifying Glasses, Barometers and Thermometers; with many other Curiosities not here mentioned.

REFRACTING TELESCOPES IMPROVED,

By a Method founded on the justest Principles and Rules of Opticks, are allowed by the best Judges in Theory (who have made the Comparison) to be equal in all Respects to those of the celebrated *Pietro Patrone* at *Milan*; by which Improvement, those of two, three, and four Feet, have by Experience been found to be much more useful at Sea, than any have hitherto been, and are made by the abovesaid *James Mann.*

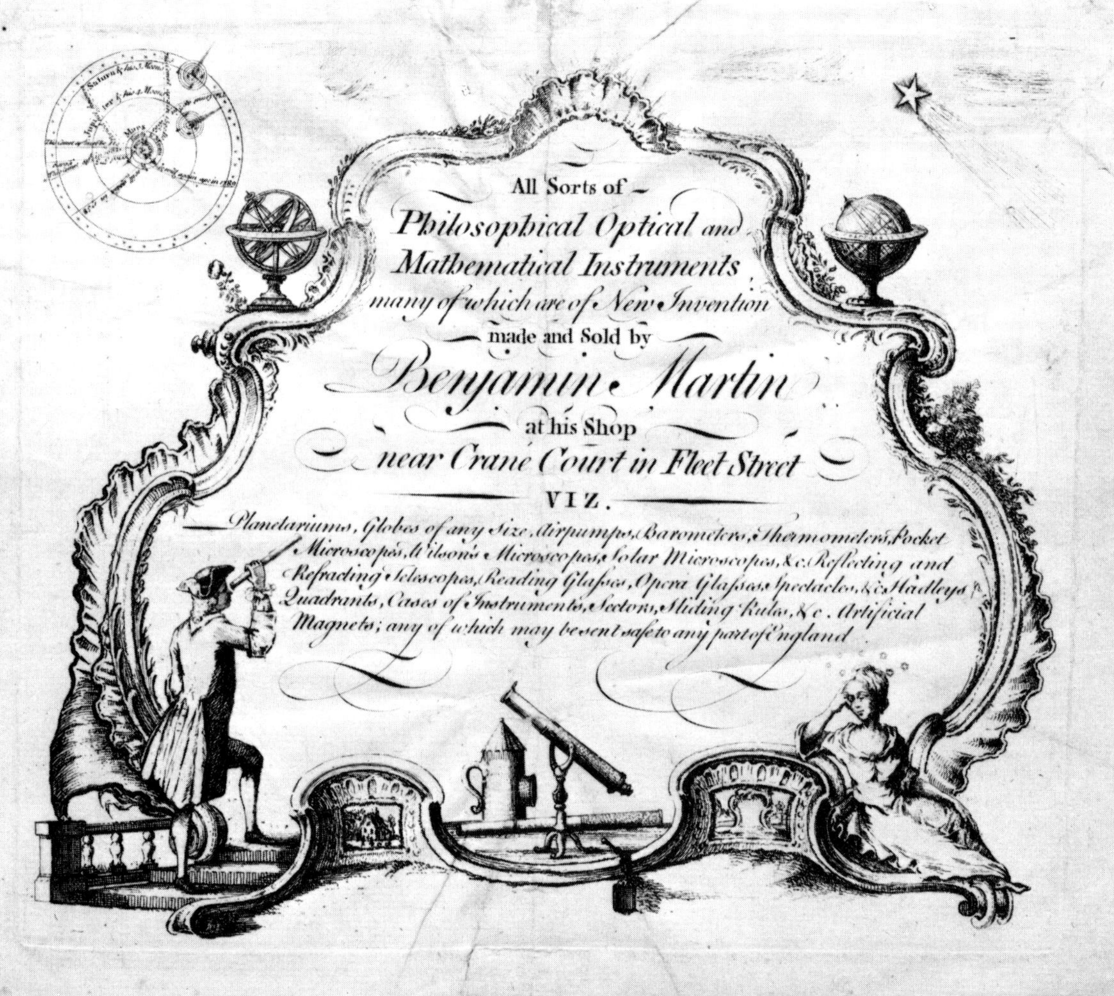

All Sorts of
Philosophical Optical and
Mathematical Instruments
many of which are of New Invention
made and Sold by
Benjamin Martin
at his Shop
near Crane Court in Fleet Street
VIZ.
Planetariums, Globes of any Size, Airpumps, Barometers, Thermometers, Pocket
Microscopes, Wilson's Microscopes, Solar Microscopes, &c. Reflecting and
Refracting Telescopes, Reading Glasses, Opera Glasses, Spectacles, &c. Hadleys
Quadrants, Cases of Instruments, Sectors, Sliding Rules, &c. Artificial
Magnets; any of which may be sent safe to any part of England

left

34 Benjamin Martin
catalogue no.258

right

35 Edward Nairn
catalogue no.281

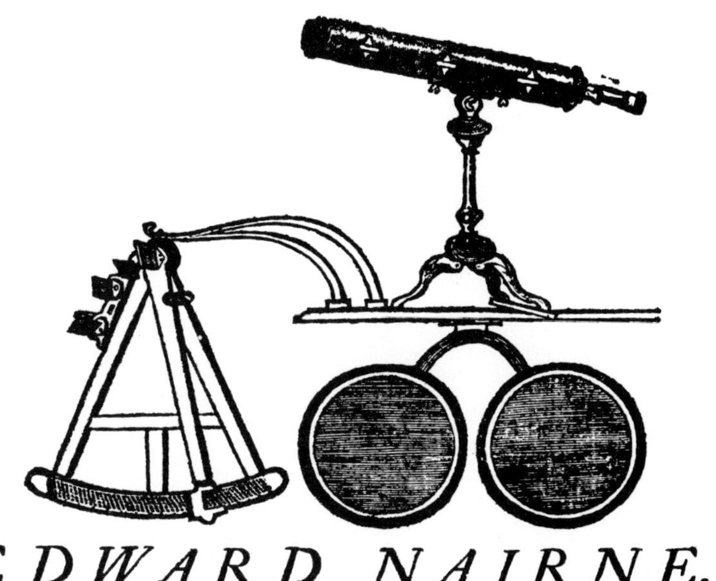

EDWARD NAIRNE,

Optical, Philofophical, *and* Mathematical Inftrument-Maker,

At the GOLDEN SPECTACLES, REFLECTING TELESCOPE *and* HADLEY'S QUADRANT, *in* Cornhill, *oppofite the* Royal Exchange, LONDON;

MAKES and Sells Spectacles, either of GLASS or BRAZIL PEBBLE, fet in neat and commodious Frames, fome of which neither prefs the Nofe nor Temples.
CONCAVES for Short-Sighted Perfons; READING, BURNING and MAGNIFYING-GLASSES.

Newtonian and *Gregorian* REFLECTING TELESCOPES; alfo EQUATORIAL TELESCOPES, or PORTABLE OBSERVATORIES.

REFRACTING TELESCOPES of all Sorts; particularly one of a new CONSTRUCTION (which may be ufed either at Land or Sea) that will ftand all Kinds of Weather without warping, and is allowed by thofe efteemed the beft JUDGES, who have made feveral late Trials of them at Sea, to exceed all others yet made in *England*; alfo a peculiar Sort to be ufed at Sea in the Night.

MICROSCOPES, either DOUBLE, SINGLE, AQUATICK, SOLAR, or OPAKE; likewife a new-invented POCKET MICROSCOPE, that may be ufed with the SOLAR, and anfwers the Purpofes of all the other Sorts.

CAMERA OBSCURAS for delineating Landfkips and Profpects (and which ferve to view Perfpective Prints) made with truly Parallel Planes; SKY-OPTIC BALLS; PRISMS for demonftrating the Theory of Light and Colours; CONCAVE, CONVEX, and CYLINDRICAL SPECULUMS; MAGICK LANTERNS; OPERA GLASSES; OPTICAL MACHINES for Perfpective Prints; CYLINDERS and CYLINDRICAL PICTURES.

AIR PUMPS, particularly a fmall Sort that ferves for Condenfing; AIR FOUNTAINS of various Kinds; GLASS PUMPS; WIND GUNS; PAPIN'S DIGESTORS; alfo a PORTABLE APPARATUS for ELECTRICAL EXPERIMENTS, which is allowed by the CURIOUS to exceed any of the Kind.

BAROMETERS, DIAGONAL, STANDARD, or PORTABLE.

THERMOMETERS, whofe Scales are adjufted to the Bores of their refpective Tubes; HYGROMETERS; HYDROSTATICAL BALANCES, and HYDROMETERS.

HADLEY'S QUADRANTS, after the moft exact Method, with Glaffes, whofe Planes are truly parellel; DAVIS'S QUADRANTS; GLOBES of all Sizes; AZIMUTH and other Sea Compaffes; LOADSTONES; NOCTURNALS; SUN-DIALS of all Sorts; Cafes of DRAWING INSTRUMENTS; SCALES; PARALLEL RULERS; PROPORTIONAL COMPASSES and DRAWING PENS.

THEODOLITES, SEMICIRCLES, CIRCUMFERENTERS, PLAIN TABLES, DRAWING BOARDS, MEASURING WHEELS, SPIRIT LEVELS, RULES, PENCILS, and all other Sorts of OPTICAL, PHILOSOPHICAL, and MATHEMATICAL INSTRUMENTS, of the neweft and moft approved Inventions, are made and fold by the abovefaid *EDWARD NAIRNE.*

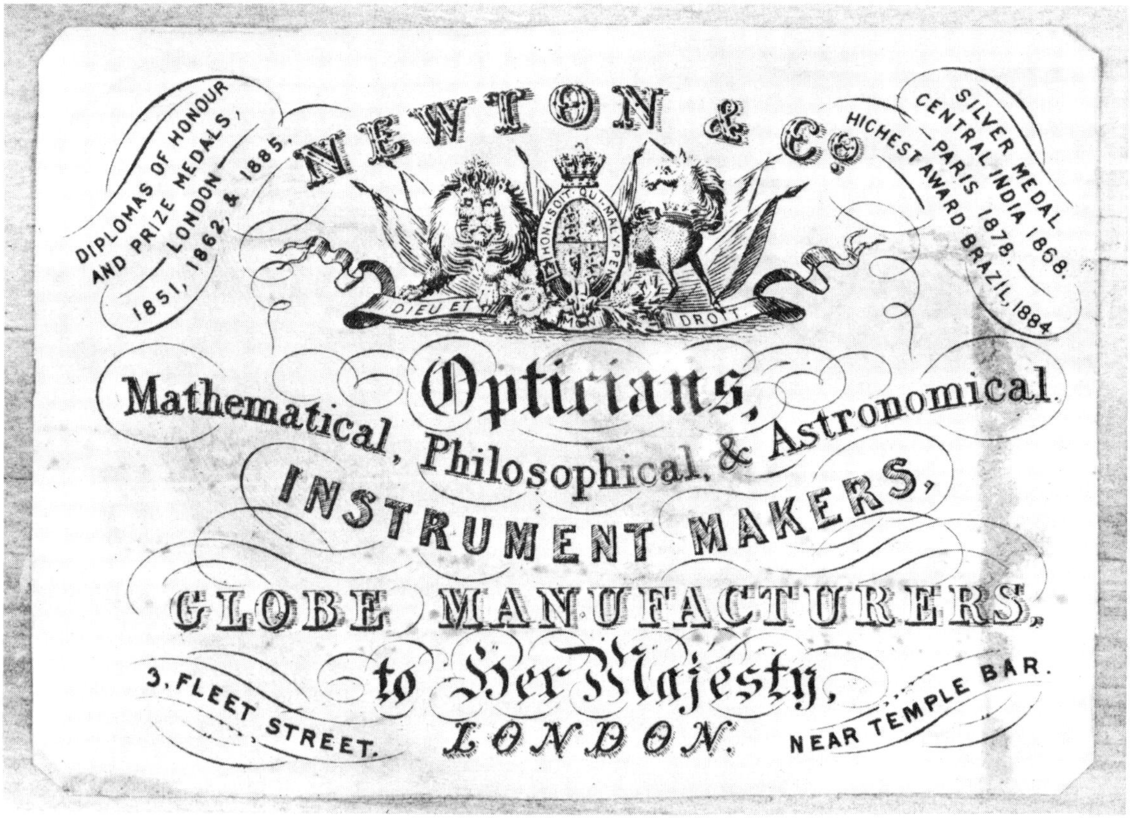

DIPLOMAS OF HONOUR AND PRIZE MEDALS, LONDON, 1851, 1862, & 1885.
SILVER MEDAL — PARIS 1868. CENTRAL INDIA 1868. HIGHEST AWARD BRAZIL, 1884.
NEWTON & Co.
HONI SOIT QUI MALY PENSE
DIEU ET MON DROIT
Opticians,
Mathematical, Philosophical, & Astronomical,
INSTRUMENT MAKERS,
GLOBE MANUFACTURERS,
to Her Majesty,
LONDON.
3, FLEET STREET.
NEAR TEMPLE BAR.

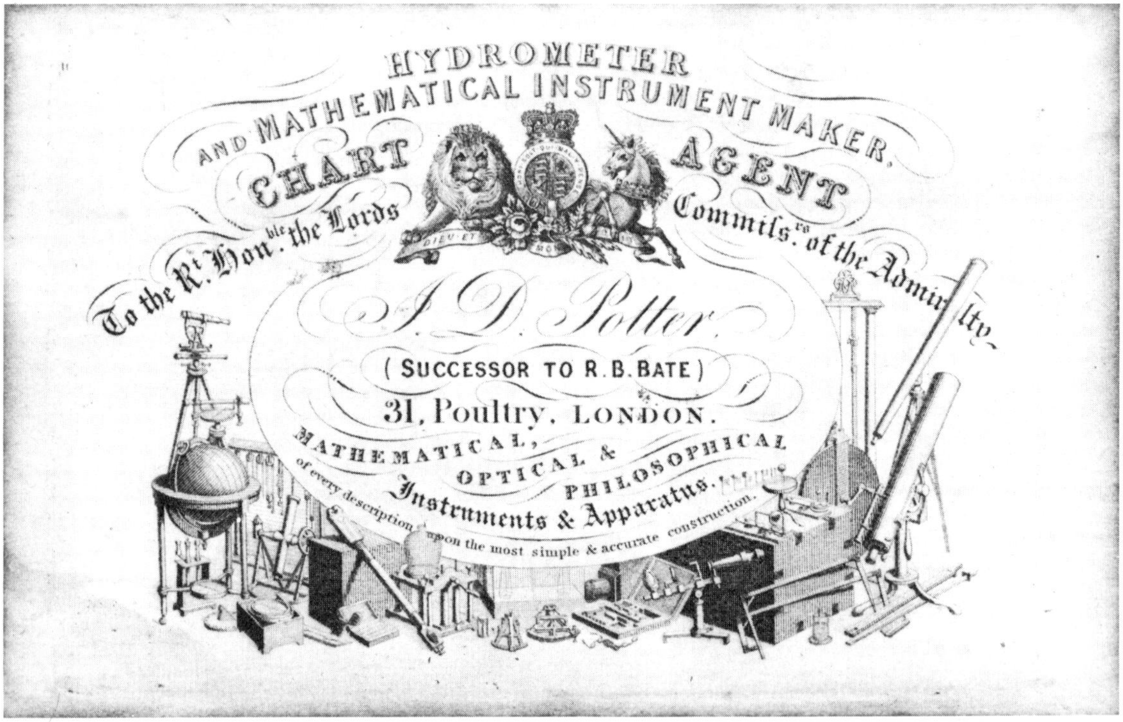

HYDROMETER
AND MATHEMATICAL INSTRUMENT MAKER,
CHART
AGENT
To the Rt. Honble. the Lords
Commifsrs. of the Admiralty.
DIEU ET MON
J. D. Potter,
(SUCCESSOR TO R. B. BATE)
31, Poultry, LONDON.
MATHEMATICAL,
OPTICAL &
PHILOSOPHICAL
Instruments & Apparatus,
of every description, upon the most simple & accurate construction.

top left

36 Newton and Co. catalogue no.288

bottom left

37 J. D. Potter catalogue no.302

right

38 Henry Pyefinch catalogue no.309

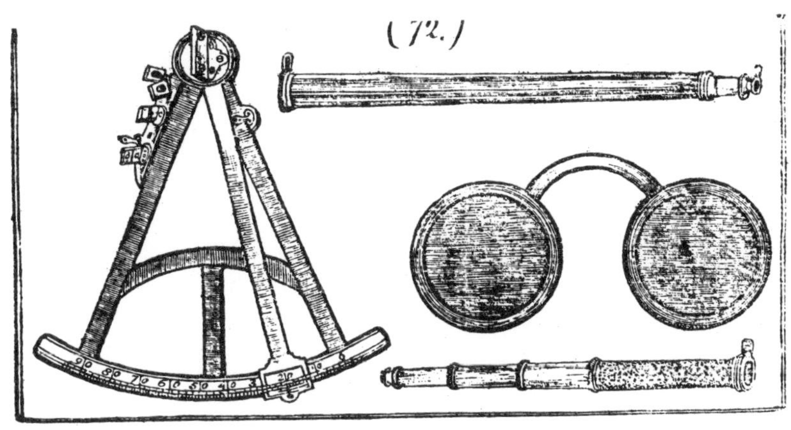

(72.)

A CATALOGUE

OF

OPTICAL, PHILOSOPHICAL, and MATHEMATICAL INSTRUMENTS,

MADE AND SOLD BY

HENRY PYEFINCH,

At the GOLDEN QUADRANT, SUN, and SPECTACLES, No. 67, between *Bishopsgate-Street*, and the *Royal-Exchange*, in *Cornhill*, *LONDON*.

	l.	s.	d.
A PAIR best silver spectacles for the temples	0	15	0
Best ditto, with double joints	1	4	0
Brazil pebbles, in silver frame	1	11	6
Ditto, in a double joint frame	1	18	0
Pebbles in best steel frame	1	1	0
Ditto in a double jointed frame	1	4	0
Best steel spectacles for the temples	0	6	6
Ditto	0	5	0
Ditto	0	2	6
Double jointed best steel	0	9	0
Best silver spectacles for nose	0	7	0
Tortoiseshell and silver	0	3	6
Spectacles for the nose	0	1	0
Spectacle cases from 1s. to	1	4	0
Concave glass for short-sighted persons, in horn	0	1	6
Same in tortoiseshell	0	3	0
In tortoiseshell and silver	0	5	6
Ditto	0	10	6
In Pearl and silver	0	12	0
Reading glasses mounted in horn, tortoiseshell, tortoiseshell and silver, &c. from 2s 6d to	1	11	6
Opera glasses, from 6s. to	1	1	0
Ditto, nourse skin and silver	1	11	6
Larger 2l. 2s. and	2	12	6
Refracting telescopes for sea and land, with achromatic moving object glasses, 20 inches	1	11	6
Two feet	2	2	0
Three feet	3	3	0
and so in proportion to any length			
Refracting telescopes, with six glasses, two feet long	0	18	0
Three feet, ditto	1	4	0
Four feet, ditto	1	11	6
Refracting telescopes, with 4 glasses, 3 feet painted tube	0	5	0
Three feet in mahogany	0	7	6
Ditto	0	10	6
Refracting telescopes for pocket, of all sorts and sizes, from 5s to	1	11	6
Reflecting telescopes, for the pocket	2	12	6
Fourteen inches long	5	5	0
Eighteen inches	7	7	0
Two feet long	10	10	0
Ditto, with rack work, finder, &c.	16	16	0
Best compound microscope	6	6	0
Ditto, with joint	7	7	0
Ditto, with three pillars	4	4	0
Small sort, ditto	2	12	6
Wilson's microscope	2	2	0
The same adapted for opake objects	2	10	0
Solar microscope compleat	5	5	0
Ditto with magelliscope	8	8	0
Opake microscope	2	2	0
Prisms, 8s. to	0	18	0
Magic lanthorn	1	1	0
Painted figures for ditto	0	5	0
Others from 7s. 6d. to	0	10	6
Concave and convex mirrors of all dimensions from 9s. to	31	10	0
Cylindrical concave	1	1	0
Ditto, larger	1	11	6
Cylindrical convex, and its six pictures	2	12	0
A pocket camera	0	10	6
Ditto, larger	0	15	0
Camera obscura, in form of a book, the best	4	4	0
The same adapted to view prints	5	5	0
Diagonal mirror, or optic machine	0	15	0
Ditto, for two people	1	11	6
Ditto, in form of book,	1	15	0
The best coloured views pasted, at per dozen	0	18	0

MATHEMATICAL INSTRUMENTS.

	l.	s.	d.
Hadley's octant	1	10	0
Ditto, ivory Ar. and Nonius	2	2	0
Ditto, brass index	2	12	6
Ditto in ebony	3	3	0
Small brass octant	2	12	6
Twelve inch ditto	5	5	0
Eighteen inch ditto edge bars	10	10	0
Davis's and all other quadrants, from 5s. to	0	12	0
Theodolites, from 5l. 5s. to	25	0	0
Circumferenter	2	10	0
Plane table, ball, socket, and staff	3	10	0
Gunter's chains and measuring tapes, of all prices			
Cases with drawing Instruments, from 7s. 6d. to	9	9	0
Protractors from 1s. 6d. to	4	4	0
Proportional compasses	1	5	0
Ditto larger	1	10	0
Pantographer in ebony	2	12	6
Pantographer, all brass	4	14	6
Gunner's callipers	1	11	6
Parallel rules, from 3s. 6d. to	2	12	6
Horizontal sun dials, 6s. to	10	10	0
Ring dial universal	0	12	0
Ditto, six inches diameter	0	18	0
Nine inches, best	5	5	0
Ditto on pedestal, very curious	20	0	0
Azimuth compasses	5	5	0
and all other sorts and inferior prices			
Artificial magnets, to lift any weight not exceeding three hundred	1	1	0
	2	2	0
	3	3	0
Perambulator, or measuring wheel	6	6	0
Astronomical quadrants, from 15l. 15s. to	150	0	0
Globes 17 inches. the pair	6	6	0
Twelve inches	3	3	0
Nine inches	2	2	0
Six inches	1	11	6
Three inches, for pockets	0	9	0

PHILOSOPHICAL INSTRUMENTS.

	l.	s.	d.
Best barometer	1	11	6
Ditto	2	2	0
Barometer with thermometer	2	12	6
The same with thermometer and hygrometer	3	3	0
Best sort	4	4	0
Fahrenheit's best thermometer	1	11	6
Ditto	1	1	0
Botannic thermometer	1	1	0
Ditto for	0	16	0
Best hydrometer, with all its weights	1	1	0
Hydrostatic balance	1	16	0
An electrical machine	3	13	6
Ditto, large	5	5	0
Electrometers for ditto	0	10	6
Air pump all brass double barrel	3	13	6
Ditto	5	15	6
Ditto, mahogany and brass, large	7	7	0
Ditto, the most compleat	9	9	0
Apparatus for ditto, from 1l. 11s. 6d to	9	9	0
A pirometer to try the expansion of metals	2	13	6

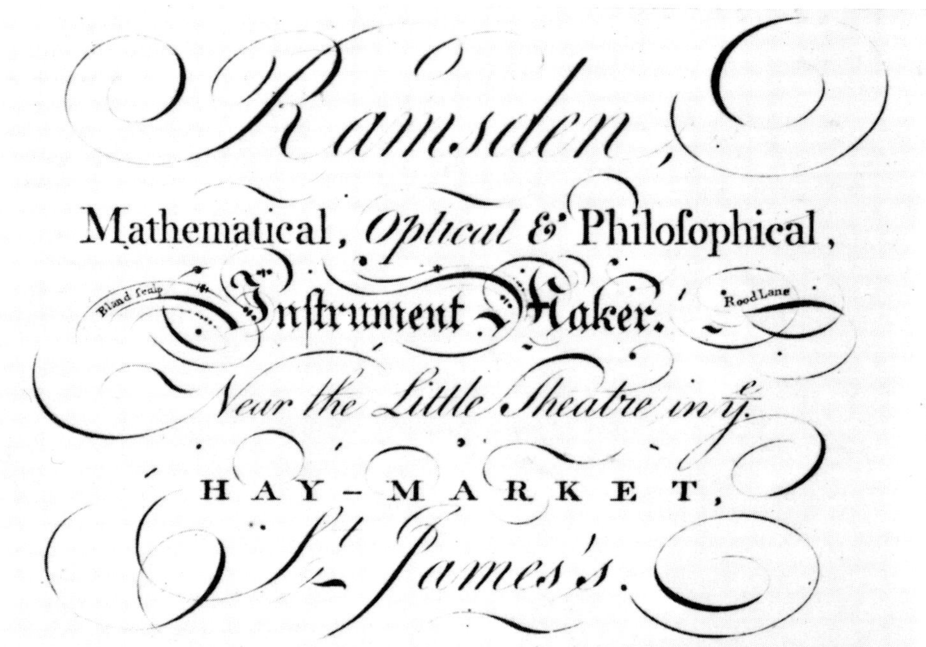

Ramsden,
Mathematical, Optical & Philosophical,
Bland sculp
Rood Lane
Instrument Maker.
Near the Little Theatre in y.
HAY-MARKET,
St James's.

left
39 Ramsden
catalogue no.311

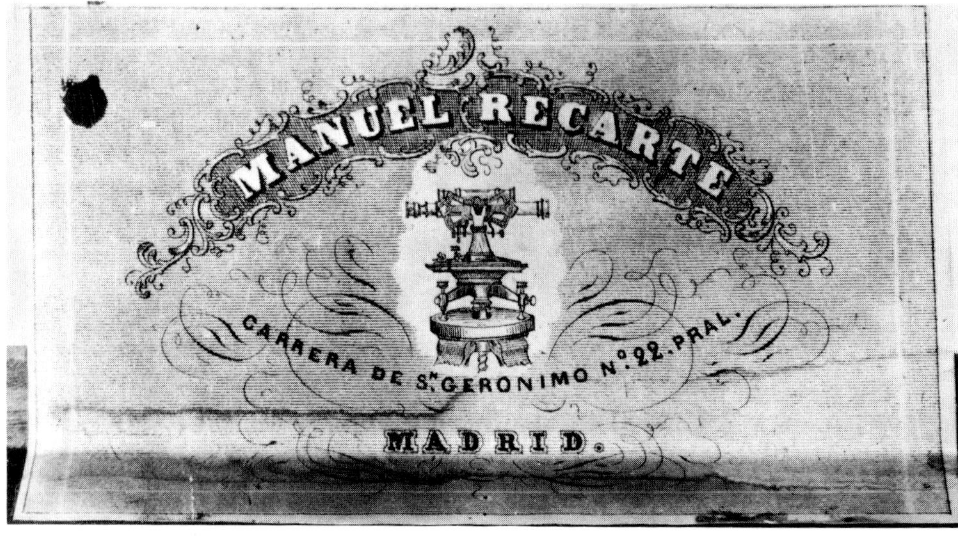

MANUEL RECARTE
CARRERA DE Sn GERONIMO No 22. PRAL.
MADRID.

left
40 Manuel Recarte
catalogue no.314

right
41 Thomas Ribright
catalogue no.316

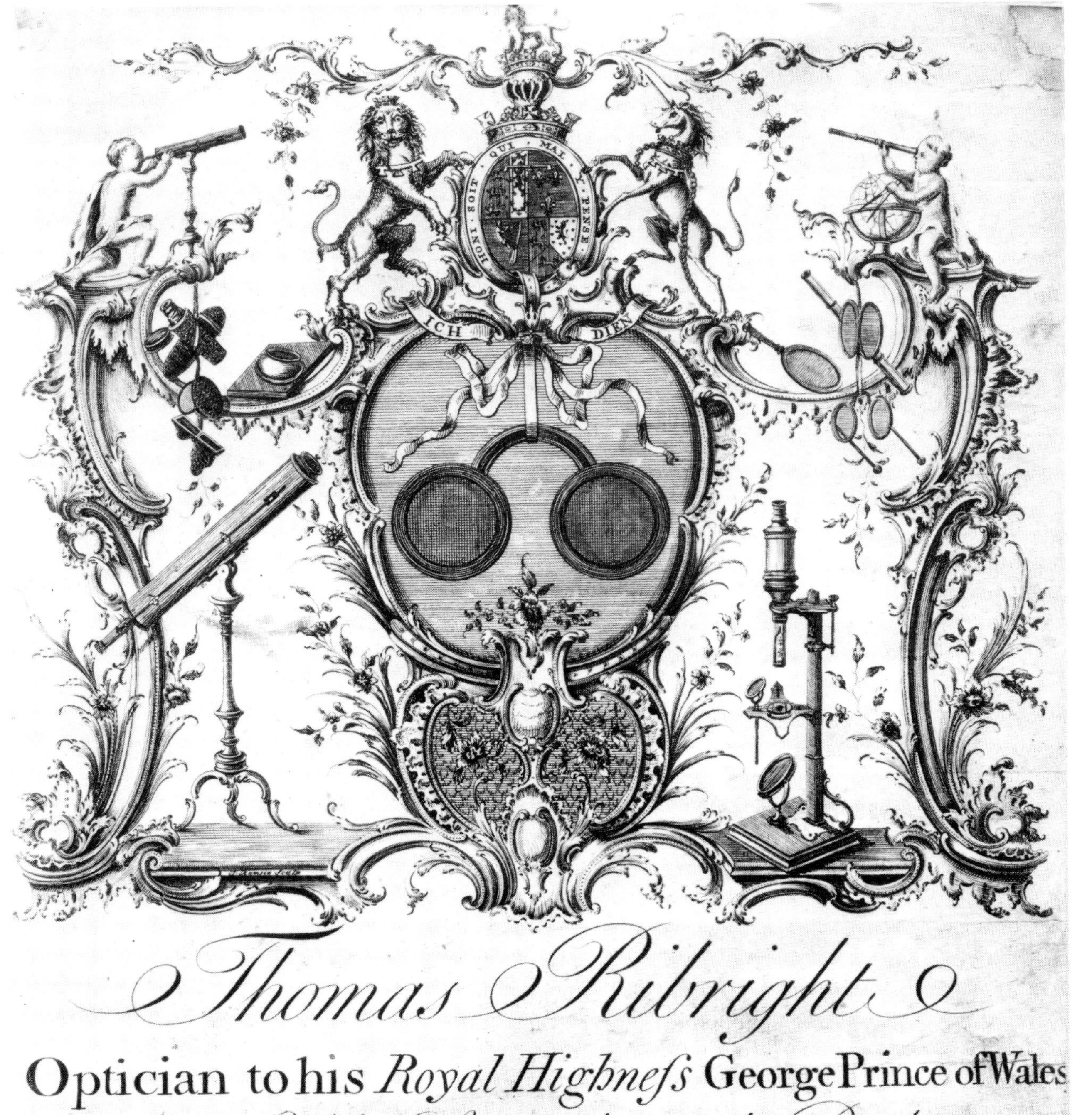
HONI SOIT QUI MAL Y PENSE
ICH DIEN
Thomas Ribright
Optician to his Royal Highness George Prince of Wales
at the Golden Spectacles in the Poultry,
London.

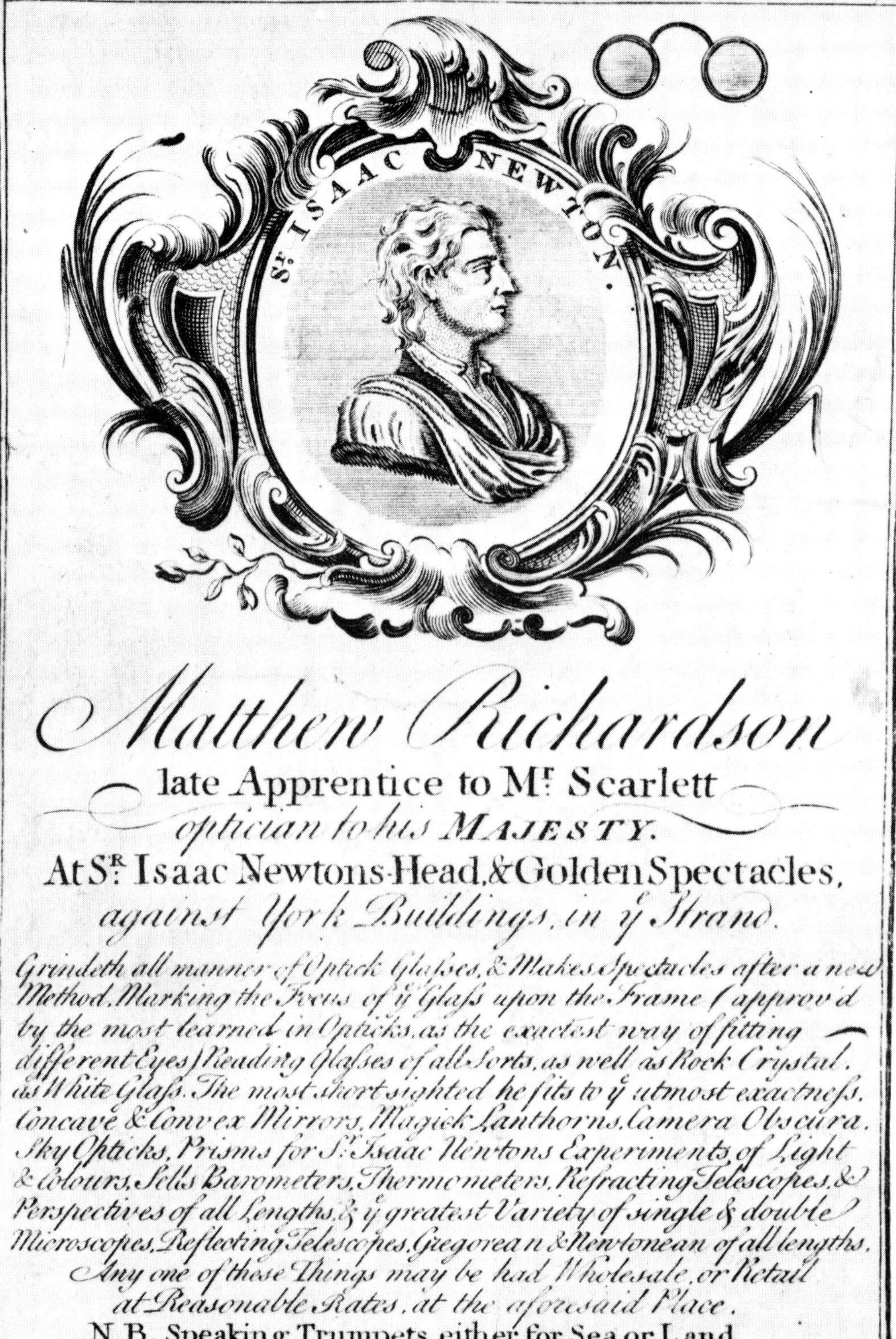

Matthew Richardson
late Apprentice to Mr. Scarlett
optician to his MAJESTY.
At Sr. Isaac Newtons Head, & Golden Spectacles,
against York Buildings, in ye Strand.

Grindeth all manner of Optick Glasses, & Makes Spectacles after a new Method, Marking the Focus of ye Glass upon the Frame (approv'd by the most learned in Opticks, as the exactest way of fitting different Eyes) Reading Glasses of all Sorts, as well as Rock Crystal, as White Glass. The most short sighted he fits to ye utmost exactness, Concave & Convex Mirrors, Magick Lanthorns, Camera Obscura, Sky Opticks, Prisms for Sr. Isaac Newtons Experiments of Light & Colours, Sells Barometers, Thermometers, Refracting Telescopes, & Perspectives of all Lengths, & ye greatest Variety of single & double Microscopes, Reflecting Telescopes, Gregorean & Newtonean of all lengths. Any one of these Things may be had Wholesale, or Retail at Reasonable Rates, at the aforesaid Place.

N.B. Speaking Trumpets either for Sea or Land.

42 Matthew Richardson
catalogue no.317

43 R. Rust
catalogue no.334

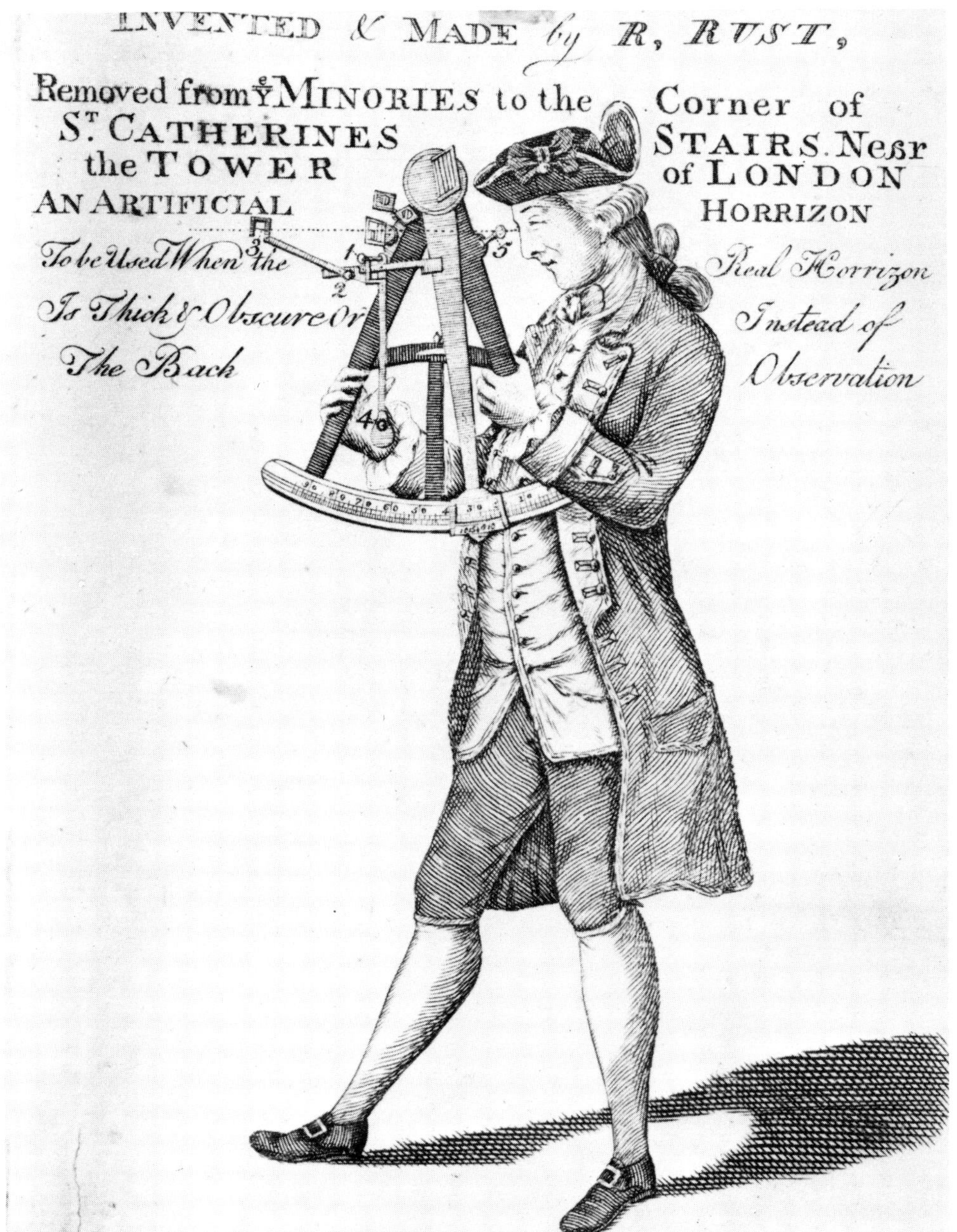

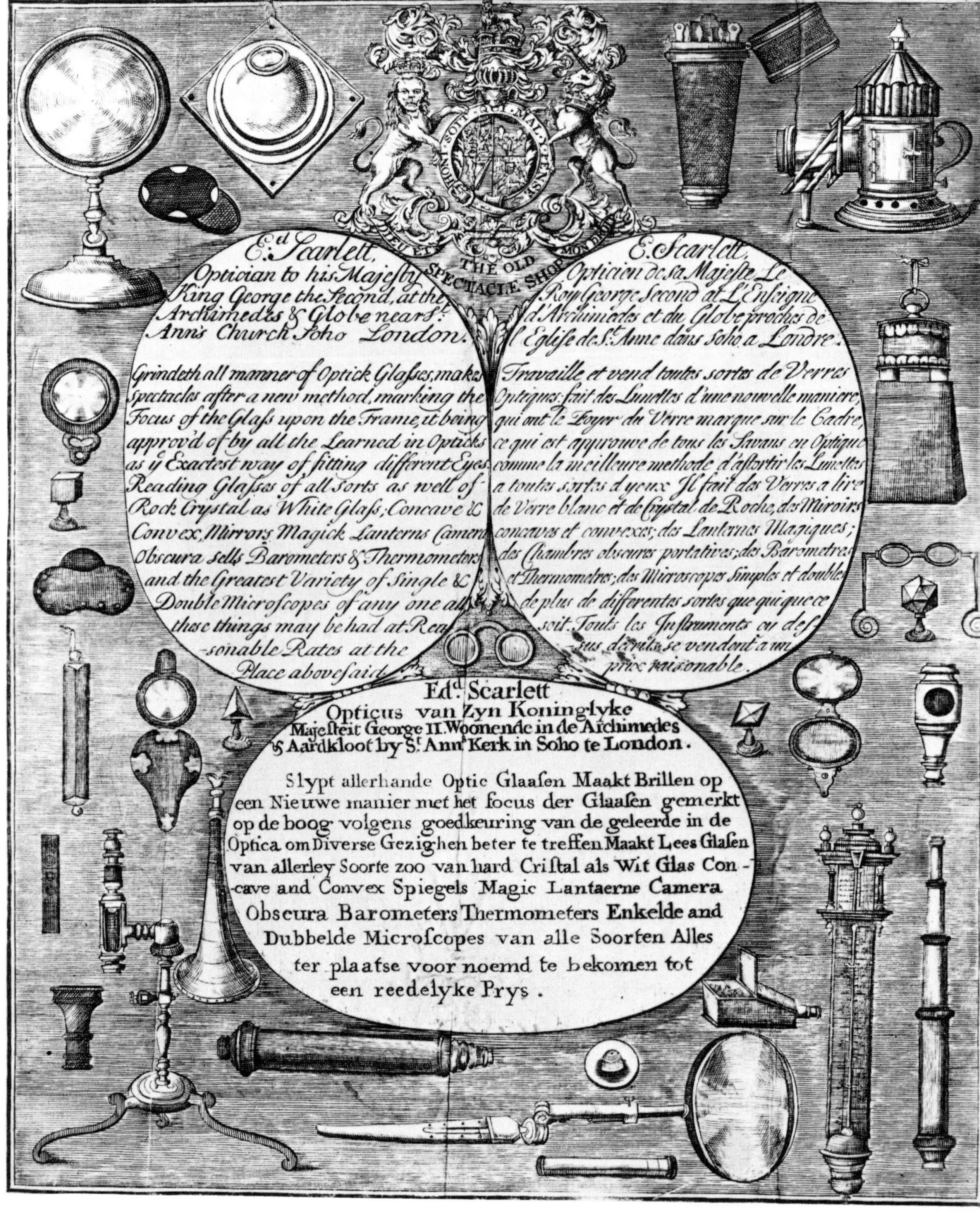
THE OLD SPECTACLE SHOP

Ed. Scarlett.
Optician to his Majesty
King George the Second, at the
Archimedes & Globe near St.
Ann's Church Soho London.

Grindeth all manner of Optick Glasses, makes
Spectacles after a new method, marking the
Focus of the Glass upon the Frame, it being
approv'd of by all the Learned in Opticks
as ye Exactest way of fitting different Eyes
Reading Glasses of all sorts as well of
Rock Crystal as White Glass; Concave &
Convex, Mirrors, Magick Lanterns, Camera
Obscura sells Barometers & Thermometers
and the Greatest Variety of Single &
Double Microscopes of any one all
these things may be had at Rea-
sonable Rates at the
Place abovesaid.

E. Scarlett.
Opticien de sa Majesté, Le
Roy George Second a l'Enseigne
d'Archimedes et du Globe proche de
l'Eglise de St. Anne dans Soho a Londre.

Travaille et vend toutes sortes de Verres
Optiques, fait des Lunettes d'une nouvelle maniere
qui ont le Foyer du Verre marque sur le Cadre
ce qui est approuvé de tous les Savans en Optique
comme la meilleure methode d'assortir les Lunettes
a toutes sortes d'yeux Il fait des Verres a lire
de Verre blanc et de Crystal de Roche, des Miroirs
concaves et convexes; des Lanternes Magiques;
des Chambres obscures portatives; des Barometres
et Thermometres; des Microscopes simples et doubles
de plus de differentes sortes que qui que ce
soit. Tous les Instruments ci des-
sus décrits se vendent a un
prix raisonable.

Ed. Scarlett
Opticus van Zyn Koninglyke
Majesteit George II. Woonende in de Archimedes
& Aardkloot by St. Ann. Kerk in Soho te London.

Slypt allerhande Optic Glaasen Maakt Brillen op
een Nieuwe manier met het focus der Glaasen gemerkt
op de boog volgens goedkeuring van de geleerde in de
Optica om Diverse Gezighen beter te treffen Maakt Lees Glasen
van allerley Soorte zoo van hard Cristal als Wit Glas Concave and Convex Spiegels Magic Lantaerne Camera
Obscura Barometers Thermometers Enkelde and
Dubbelde Microscopes van alle Soorten Alles
ter plaatse voor noemd te bekomen tot
een reedelyke Prys.

left
44 Edward Scarlett
catalogue no.339

below
45 Samuel Scatliff
catalogue no.340

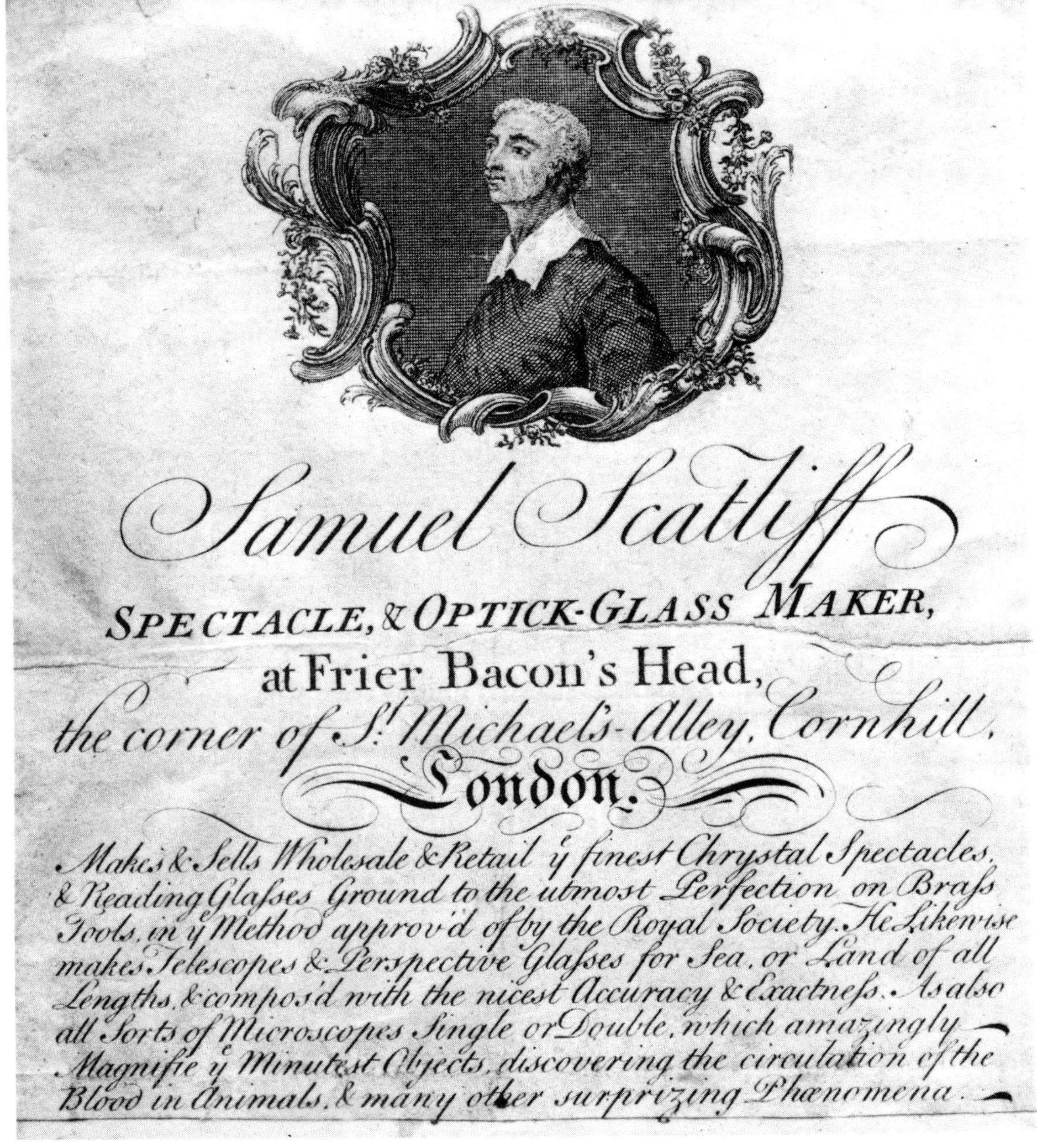

Samuel Scatliff
SPECTACLE, & OPTICK-GLASS MAKER,
at Frier Bacon's Head,
the corner of S.t Michael's-Alley, Cornhill,
London.
Makes & Sells Wholesale & Retail ÿ finest Chrystal Spectacles,
& Reading Glasses Ground to the utmost Perfection on Brass
Tools, in ÿ Method approv'd of by the Royal Society. He Likewise
makes Telescopes & Perspective Glasses for Sea, or Land of all
Lengths, & compos'd with the nicest Accuracy & Exactness. As also
all Sorts of Microscopes Single or Double, which amazingly
Magnifie ÿ Minutest Objects, discovering the circulation of the
Blood in Animals, & many other surprizing Phænomena.

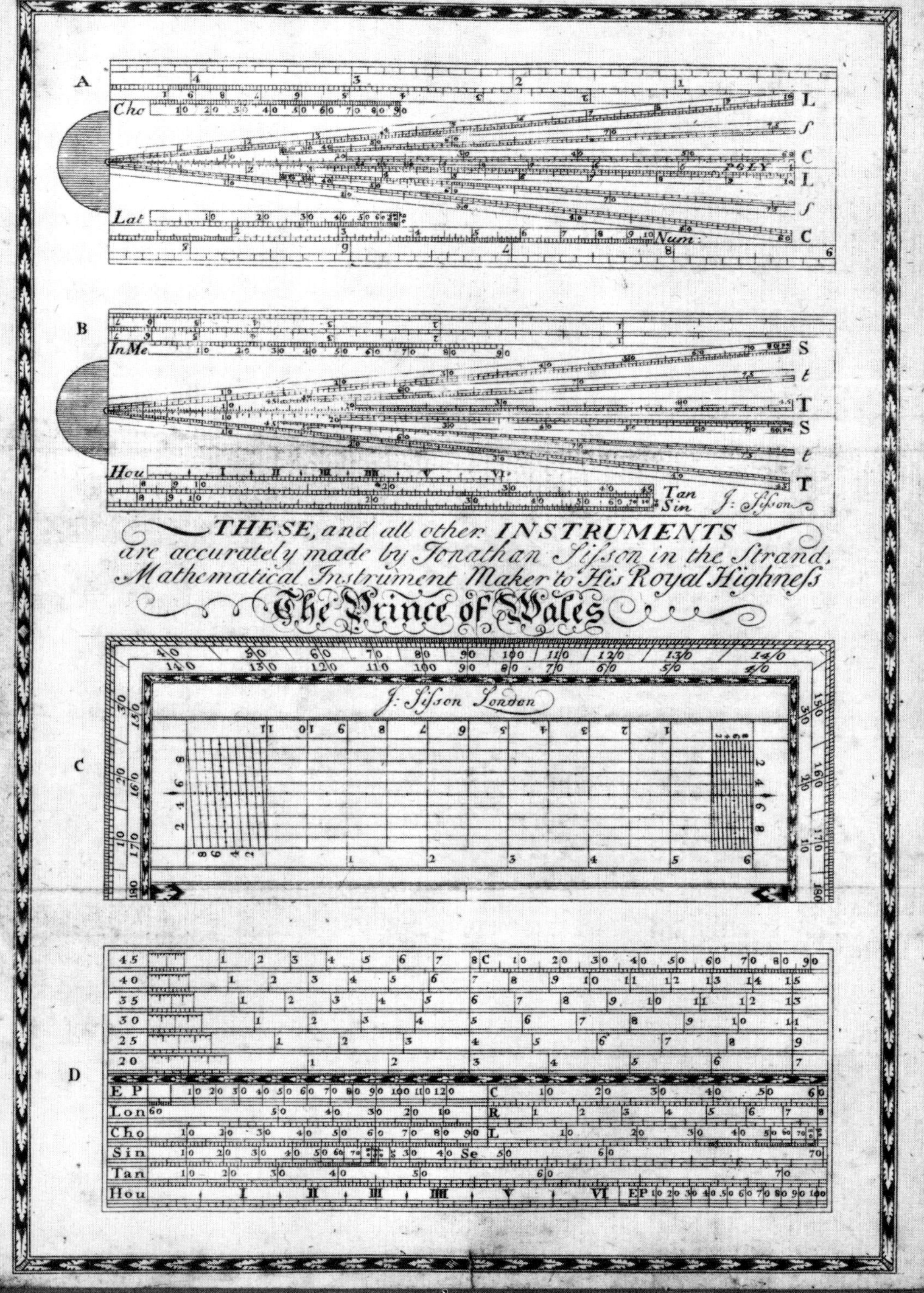
A
Cho
Lat
Num:
L
S
C
L
S
C
POLY
B
InMe
Hou
S
t
T
S
t
T
Tan
Sin
J: Sisson
THESE, and all other INSTRUMENTS
are accurately made by Jonathan Sisson in the Strand,
Mathematical Instrument Maker to His Royal Highness
The Prince of Wales
J: Sisson London
C
D
EP
Lon
Cho
Sin
Tan
Hou
C
R
L
Se
EP
I
II
III
IIII
V
VI

left

46 Jonathan Sisson
catalogue no.359

below

47 Christopher Stedman
catalogue no.381

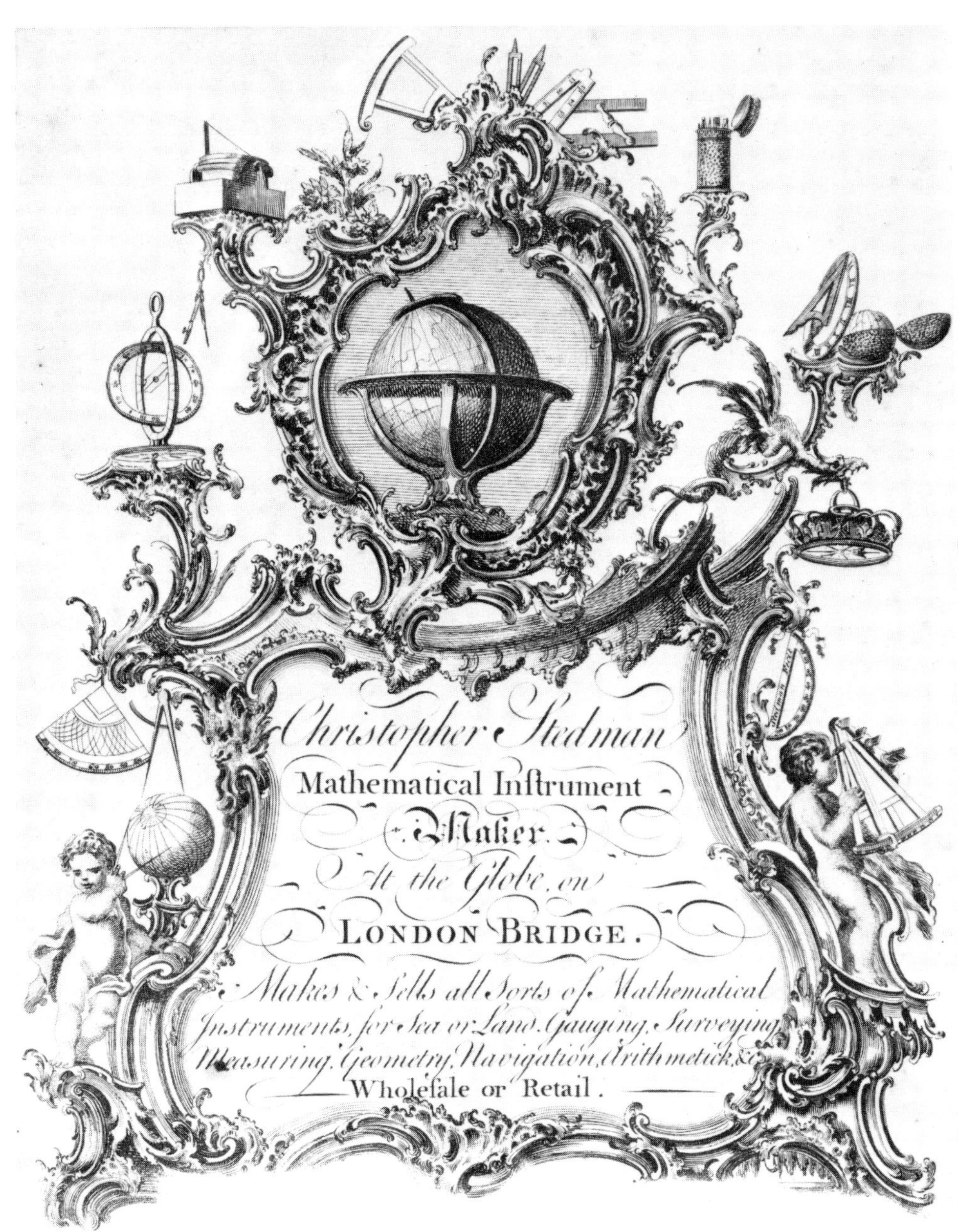

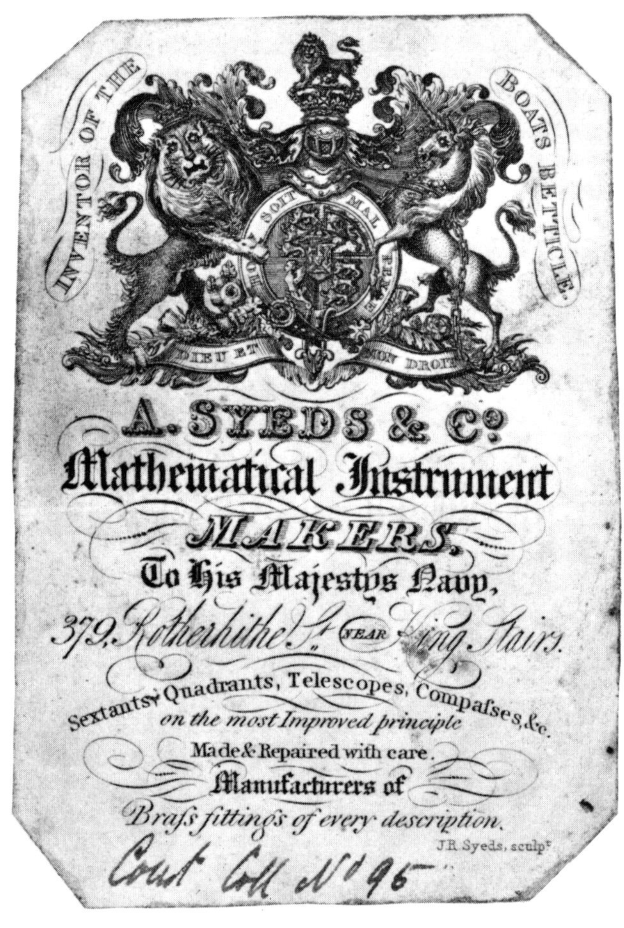

left

48 A. Syeds and Co.
catalogue no.391

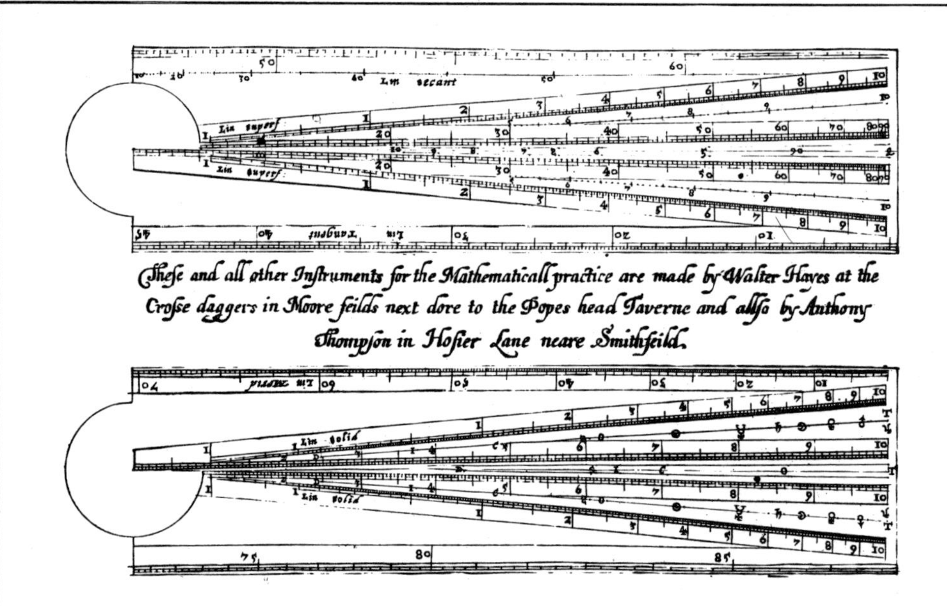

left

49 Anthony Thompson
catalogue no.395

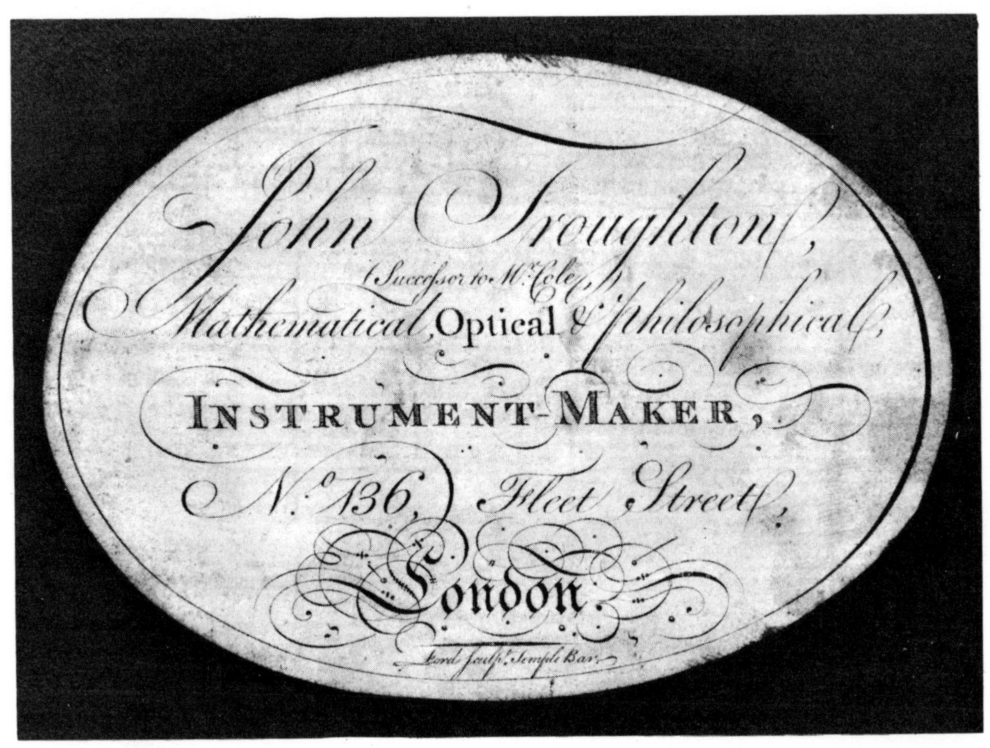

right
50 John Troughton
catalogue no.402

right
51 J. and E. Troughton
catalogue no.404

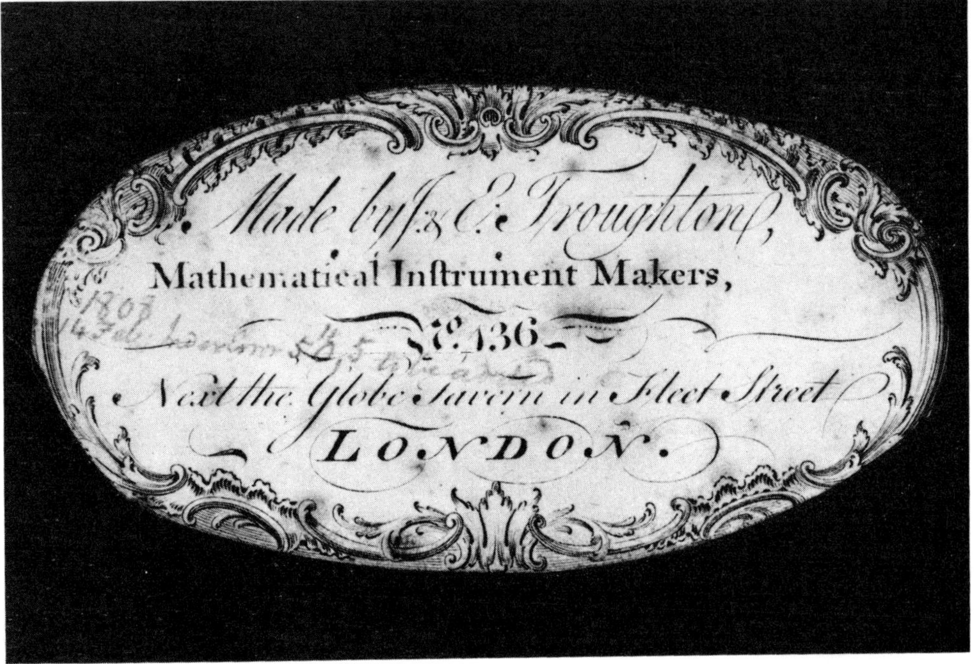

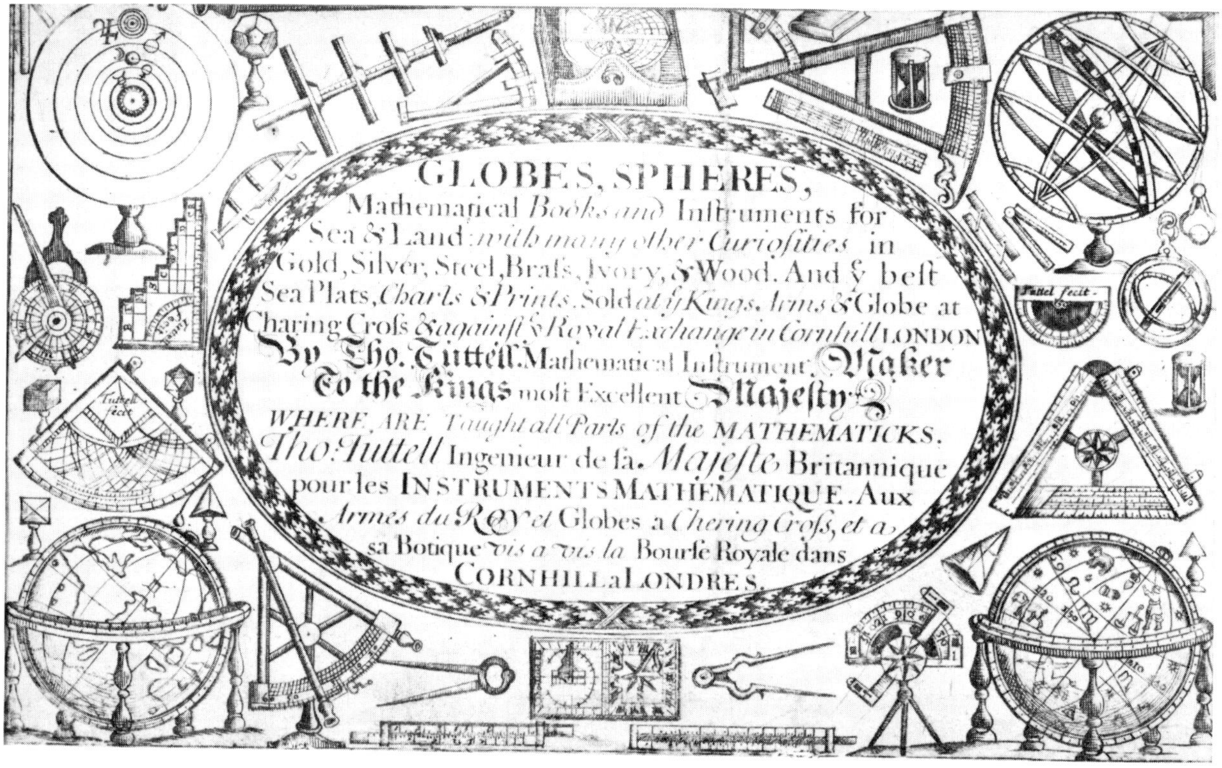

GLOBES, SPHERES,
Mathematical Books and Instruments for
Sea & Land with many other Curiosities in
Gold, Silver, Steel, Brass, Ivory, & Wood. And ye best
Sea Plats, Charts & Prints. Sold at ye Kings Arms & Globe at
Charing Cross & against ye Royal Exchange in Cornhill LONDON
By Tho. Tuttell Mathematical Instrument Maker
To the Kings most Excellent Majesty
WHERE ARE Taught all Parts of the MATHEMATICKS.
Tho: Tuttell Ingenieur de sa Majeste Britannique
pour les INSTRUMENTS MATHEMATIQUE. Aux
Armes du Roy et Globes a Chering Cross, et a
sa Botique vis a vis la Bourse Royale dans
CORNHILL a LONDRES.
Tuttel fecit

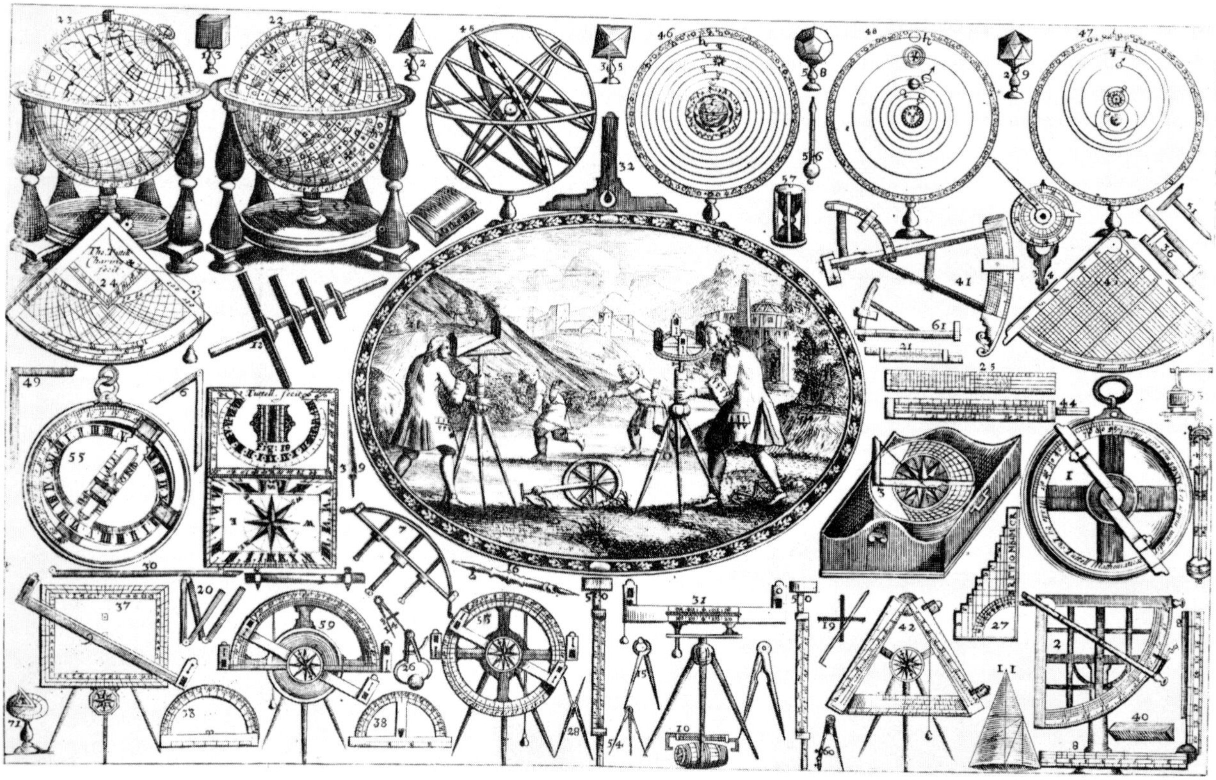

left

52 Thomas Tuttell
catalogue no.407

bottom left

53 Thomas Tuttell
catalogue no.408

top right

54 Thomas Tuttell
catalogue no.410

bottom right

55 John Walsh Walsh
catalogue no.424

Globes Cœleſtial and Terreſtrial of ſeveral Sizes, alſo in Caſes for the Pocket, with *Spheres* according to the ſeveral *Syſtemes* and all other Mathematical Inſtruments curiouſly made (at Reaſonable Rates) by *Tho. Tuttell* at the *Kings-Arms* and *Globe* at *Charing-Croſs*, and againſt the *Royal Exchange* in *Cornhill*, *London*, Mathematical Inſtrument-maker to the KING's moſt excellent Majeſty.

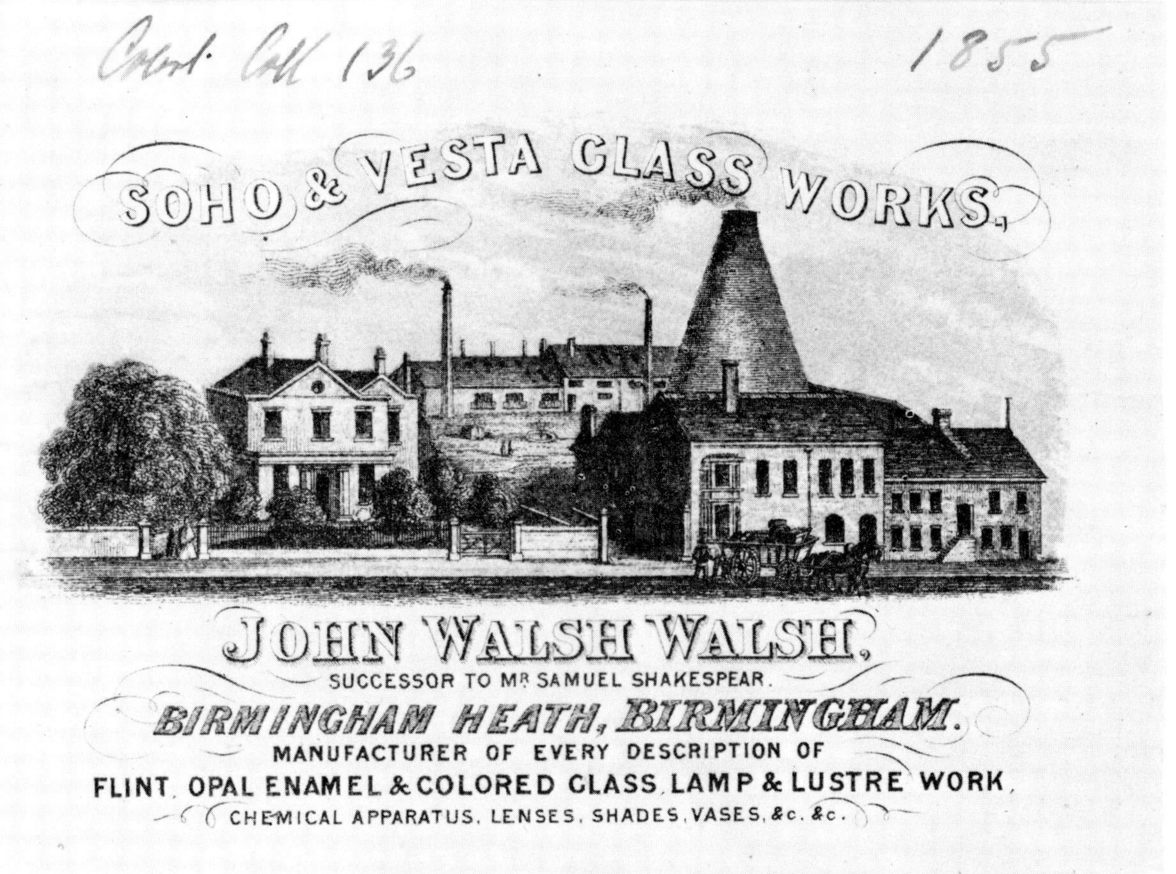

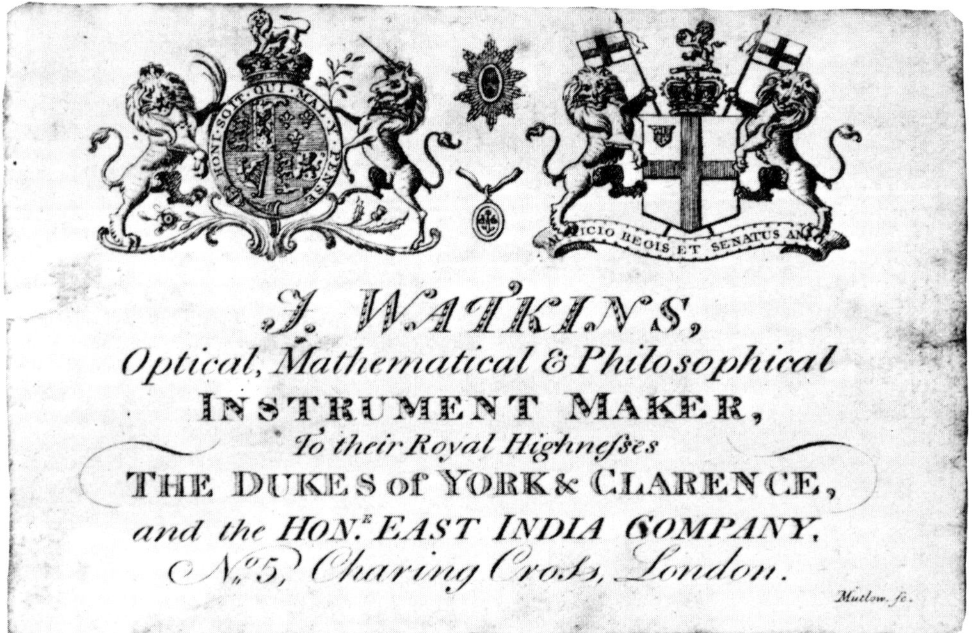

ADVERTISEMENT.

ALL the above-named Instruments as Telescopes of all Lengths, Microscopes single and double, Perspectives great and small, Reading Glasses of all sizes, Magnifying Glasses, Multiplying Glasses, Triangular Prisms, Speaking Trumpets, Spectacles fitted to all Ages, and all other Sorts of Glasses, both Concave and Convex are made and sold by JOHN YARWELL at the Archimedes and Three Golden Prospects, near the great North-Door in S. Paul's Church-Yard: London.

SAMUEL WHITFORD,

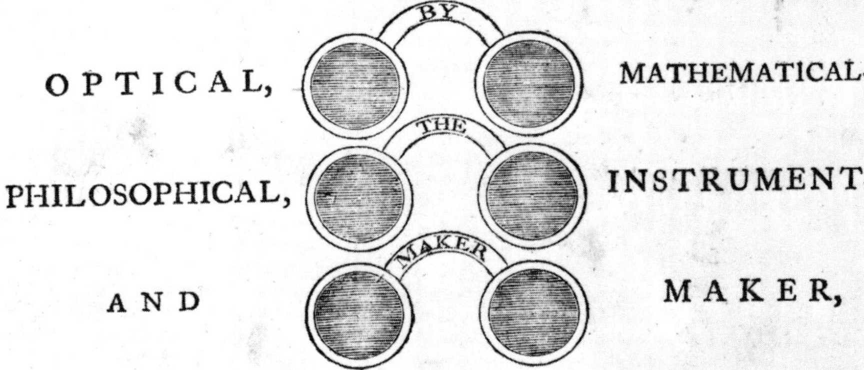

OPTICAL, MATHEMATICAL-

PHILOSOPHICAL, INSTRUMENT-

AND MAKER,

(THE ORIGINAL SHOP.)

At the THREE SPECTACLES, N.º 27,
Ludgate-Street, near *St. Paul's*, LONDON,

MAKES and sells Spectacles of Glass or Pebble, in neat and light Frames, of the newest and best Construction, and in the most convenient Manner for avoiding Pressure on the Nose or Temples.

Concaves for very short-sighted Persons; and Glasses for reading, magnifying, and burning.

Reflecting Telescopes, of all Kinds, and with the latest Improvements.

Refracting Telescopes of all Sorts; with one of a New Invention (useful for Sea or Land) that will not warp in any Weather; and a Sort to use at sea in the Night.

Double, Single, Solar, Opake, or Aquatic Microscopes.

Camera Obscuras, to delineate Landscapes and Prospects (and which serve to view Perspective Prints) made truly parallel; Sky-Optic Balls; Prisms, to demonstrate the Theory of Light and Colours; Concave, Convex, and Cylindrical Speculums; Magical Lanthorns; Opera Glasses; Optical Machines for Perspective Prints; Cylinders and Cylindrical Pictures.

Air-Pumps and Air-Fountains of various Kinds; Glass-Pumps; Portable Apparatus for Electrical Experiments, which is allowed by the Curious to be the best of the Kind.

Barometers, Diagonal, Standard, or Portable.

Thermometers, Hydrostatical Balances, and Hydrometers.

Hadley's Quadrant, after the most exact Method, with Glasses truly parallel; Davis's Quadrant; Globes of all Sizes; Compasses, Azimuths, Steering, and Plain; Loadstones; Nocturnal and Sun-dials of all Sorts.

Scales; Cases of Drawing-Instruments; Parallel Rulers; Proportional Compasses, and Drawing Pens.

Theodolites, Semicircles, Circumferenters, Measuring Wheels, Spirit Levels, Rules, and all Sorts of the best Black Lead Pencils, and all other Sorts of Instruments of the newest and most approved Invention.

** LANDS accurately surveyed.

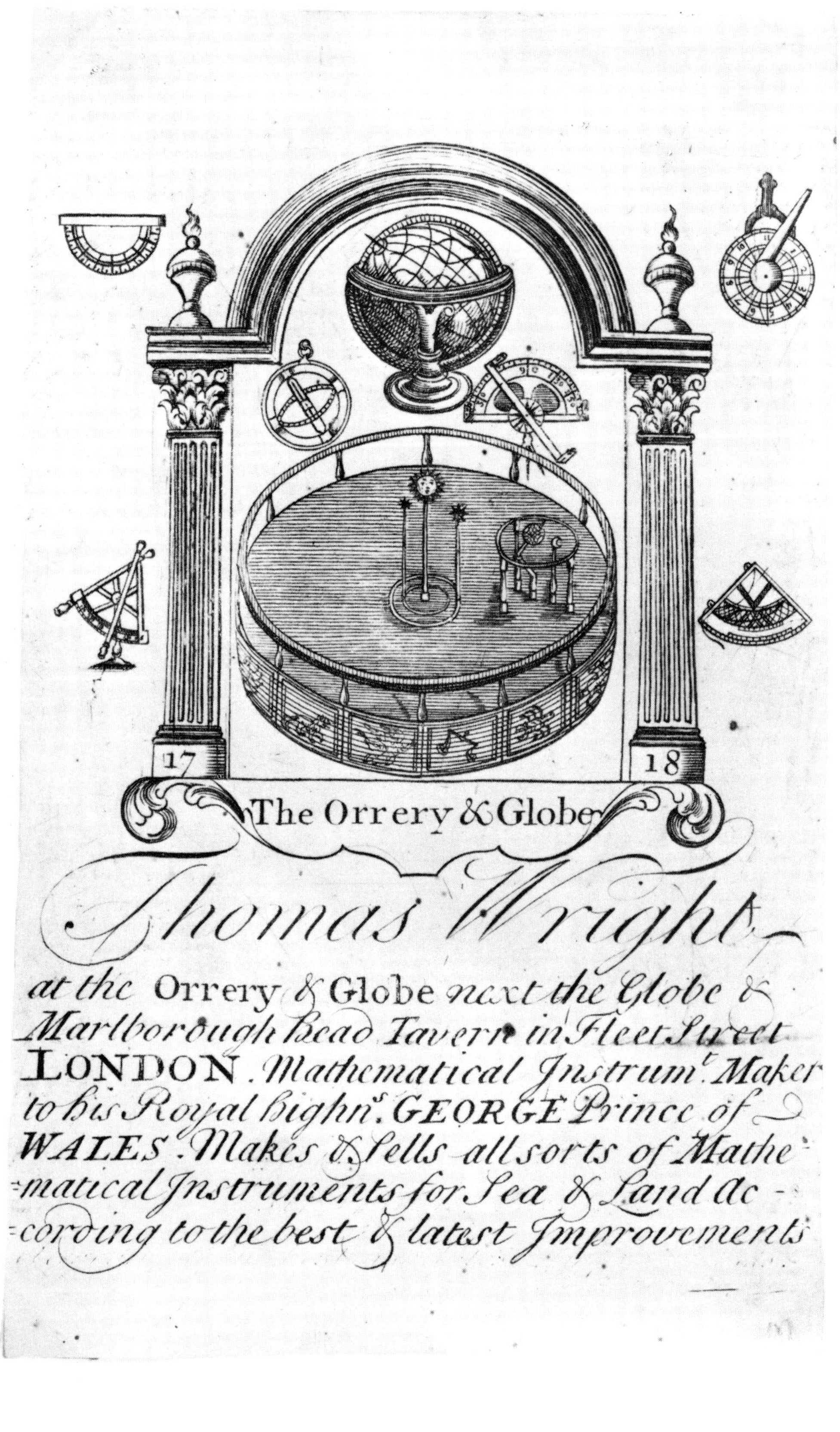

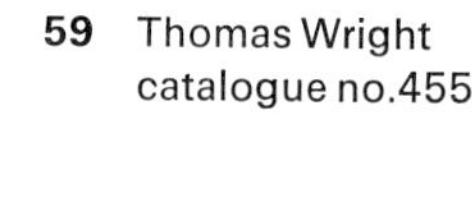

59 Thomas Wright
catalogue no.455

True Spectacles.

EXactly Ground on Brass Tools, by John Yarwell, Servant to His Majesty, approved on by the *Royal Society*, acknowledg'd by the best skill'd in Opticks to be ground to the greatest Perfection. Set neatly in Leather, Horn, Silver, or Tortoise-Shell Frames. Telescopes of all lengths for Day and Night; Perspectives great and small; a new double Microscope invented by the said Yarwell, fitted for all Uses, particularly that admirable Curiosity of seeing the Circulation of the Blood in small Fish, and Thousands of living Creatures in a drop of Pepper Water; Magnifying, Multiplying, and Weather-Glasses; Prisms, Concave Metal and Concave Burning-Glasses, Speaking-Trumpets, Reading-Glasses of all sizes; with all other sorts of Glasses both Concave and Convex, of the newest and most useful Invention.

All the above named Instruments are Made and Sold by John Yarwell, *who hath lived many Years in* Sr. Paul's Church-Yard, *is now removed to the Sign of* Archimedes *in* Ludgate-street, *the first Spectacle-Shop next* Ludgate.

By John Yarwell

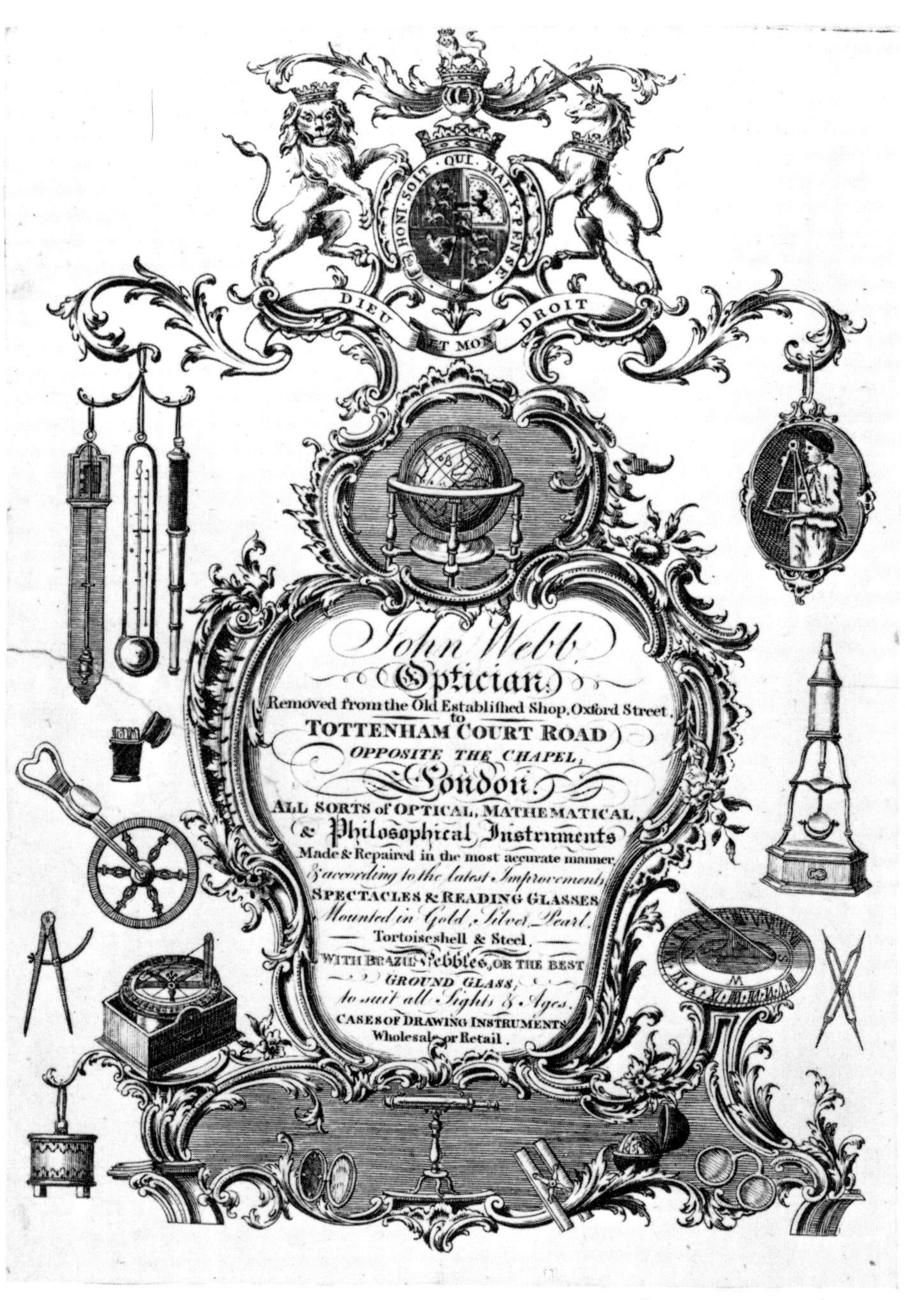
HONI SOIT QUI MAL Y PENSE
DIEU ET MON DROIT
John Webb,
Optician,
Removed from the Old Established Shop, Oxford Street,
to
TOTTENHAM COURT ROAD
OPPOSITE THE CHAPEL,
London.
ALL SORTS of OPTICAL, MATHEMATICAL,
& Philosophical Instruments
Made & Repaired in the most accurate manner,
& according to the latest Improvements,
SPECTACLES & READING GLASSES
Mounted in Gold, Silver, Pearl,
Tortoiseshell & Steel,
WITH BRAZIL Pebbles, OR THE BEST
GROUND GLASS,
to suit all Sights & Ages,
CASES OF DRAWING INSTRUMENTS
Wholesale or Retail.